Durability of
Concrete Structures

Durability of Concrete Structures

Investigation, repair, protection

Edited by
GEOFF MAYS

School of Mechanical, Materials and Civil Engineering
RMCS (Cranfield)

CRC Press
Taylor & Francis Group
Boca Raton London New York

CRC Press is an imprint of the
Taylor & Francis Group, an **informa** business

A TAYLOR & FRANCIS BOOK

CRC Press
Taylor & Francis Group
6000 Broken Sound Parkway NW, Suite 300
Boca Raton, FL 33487-2742

First issued in paperback 2019

ISBN-13: 978-0-419-15620-8 (hbk)
ISBN-13: 978-0-367-86633-4 (pbk)

Typeset in 10 on 12pt .Times by
Mews Photosetting, Beckenham, Kent

A catalogue record for this book is available from the British Library

Library of Congress Cataloging-in-Publication data available

Publisher's Note
The publisher has gone to great lengths to ensure the quality of this reprint but points out that some imperfections in the original may be apparent

Visit the Taylor & Francis Web site at
http://www.taylorandfrancis.com

and the CRC Press Web site at
http://www.crcpress.com

Contents

Preface   xi
Biographic summary of contributors   xii

PART ONE THEORIES AND SOLUTIONS

1 The behaviour of concrete   **3**
G.C. Mays

1.1 Historical development of concrete   3
1.2 Concrete — the material   4
1.3 Mechanisms and causes of failure   5
1.4 Intrinsic deterioration of concrete   6
1.5 Environmental attack   7
1.6 Prevention and repair   8
1.7 The extent of the problem   9

2 Deterioration mechanisms   **10**
T.P. Lees

2.1 Introduction   10
   2.1.1 Cement   11
   2.1.2 Hydration of cement   12
   2.1.3 Physical structure of cement paste   13
   2.1.4 Simplified model of concrete   14
2.2 Deterioration of concrete   15
   2.2.1 Early deterioration   15
   2.2.2 Deterioration arising from concrete ingredients 21
   2.2.3 Concrete deterioration caused by external agents 23
   2.2.4 External effects leading to corrosion
       of reinforcement   33

3 Structural investigations   **37**
A.F. Baker

3.1 Introduction   37

3.2 Planning an investigation 38
   3.2.1 Preliminary considerations 38
3.3 Test and inspection techniques 41
   3.3.1 *In situ* sampling and testing 41
   3.3.2 Determination of structural integrity 44
   3.3.3 Determination of concrete quality 51
   3.3.4 Steel serviceability and condition 62
   3.3.5 Laboratory testing and sample analysis 65
   3.3.6 Visual examination of concrete 66
   3.3.7 Physical testing of concrete to determine quality 67
   3.3.8 Chemical analysis of concrete 70
   3.3.9 Petrographic examination of concrete 72
3.4 Interpretation of condition surveys 76

4 Repair materials and techniques **82**
L.J. Tabor

4.1 Introduction 82
4.2 Materials for concrete repair 84
   4.2.1 Sources 84
   4.2.2 The specialist formulator 84
   4.2.3 Replacement materials – the choice 85
   4.2.4 Resin mortars 86
   4.2.5 Cementitious repair packages: examining the systems 89
   4.2.6 Crack repairs 101
   4.2.7 Superfluid microconcretes 103
4.3 Repair techniques 105
   4.3.1 Removal of defective concrete 105
   4.3.2 Cleaning the steel reinforcement 106
   4.3.3 Protecting the steel reinforcement 107
   4.3.4 Using a bond coat 108
   4.3.5 Mixing mortars and concretes 110
   4.3.6 Placing 113
   4.3.7 Crack injection 119
   4.3.8 Sprayed concrete 125
4.4 Safety 127

5 Protection **130**
A. Cathodic protection **130**
K.G.C. Berkeley

5.1 Introduction to cathodic protection 130
   5.1.1 The background to corrosion attack 130
   5.1.2 Electrochemical corrosion 131

5.2   How cathodic protection works   134
5.3   Associated problems   138
5.4   Practical applications of cathodic protection   140
5.5   Concluding remarks   145

B. Surface treatments
J.G. Keer

5.6   Introduction to surface treatments   146
5.7   Types of surface treatment   146
      5.7.1 Coatings and sealers   147
      5.7.2 Pore-lining treatments   149
      5.7.3 Pore-blocking treatments   151
      5.7.4 Renderings   152
5.8   Application of surface treatments   152
5.9   Surface treatments to control carbonation   153
5.10 Surface treatments to control moisture or chloride ingress   157
5.11 Concluding remarks   163

6. Conclusions   **166**
G.C. Mays

6.1   The future   166
6.2   Maintenance policies   167

PART TWO CASE STUDIES
7 High-rise buildings   **173**
C. Stroud and P. Gardner

7.1   Introduction   173
7.2   Five year Tall Block Inspection Programme   174
      7.2.1 Evolution of the inspection programme   174
      7.2.2 The client brief for the tall block inspection   174
      7.2.3 Methodology   175
      7.2.4 Development of the tall block inspection brief   176
      7.2.5 Typical structural defects identified by the
            inspection programme   177
7.3   High-rise repair strategies   177
      7.3.1 Public safety (emergency) works   177
      7.3.2 Holding repairs   178
      7.3.3 Full refurbishment   179
      7.3.4 Demolition   180
      7.3.5 Sale   180
7.4   Repair and maintenance strategies   180
      7.4.1 Preventative maintenance   180
      7.4.2 Criteria for strategies   181

7.5	Case studies	182
	7.5.1 Thomas North Terrace	182
	7.5.2 Herbert Willig House	185
7.6	Concluding remarks	186

8. Highway bridges — **188**
K.J. Boam

8.1	Introduction	188
8.2	Scale of the problem	188
	8.2.1 Bridge decks	189
	8.2.2 Bridge sub-structures	191
	8.2.3 Retaining walls	192
8.3	Diagnosis	192
8.4	Options for repair	194
	8.4.1 Do nothing	195
	8.4.2 Reduce corrosion rate	195
	8.4.3 Repair visual defects	195
	8.4.4 Carry out major repairs	196
	8.4.5 Apply cathodic protection	198
	8.4.6 Replace affected elements	200
8.5	Monitoring of repairs	201
8.6	Avoiding future problems	206

9 Railway structures — **206**
J.F. Dixon

9.1	Introduction	206
9.2	Steel reinforcement corrosion	207
	9.2.1 Chloride problems	207
	9.2.2 Carbonation	219
9.3	Concrete deterioration	220
	9.3.1 Frost attack	220
	9.3.2 Alkali-silica reaction	221
	9.3.3 Surface appearance	222
9.4	The future	224

10 Concrete pavements — **226**
G.D.S. Northcott

10.1	Introduction	226
10.2	Defects	227
	10.2.1 Investigation	227
	10.2.2 Types of defect	228

        10.2.3 Joint defects   228
        10.2.4 Surface defects   230
        10.2.5 Structural defects   232
        10.2.6 Material defects   232
    10.3 Design faults and remedies   234
        10.3.1 Joint seals   234
        10.3.2 Deep spalling at joints   235
        10.3.3 Longitudinal cracking   236
        10.3.4 Cracks at manholes and gallies   236
    10.4 Construction faults and remedies   237
        10.4.1 Joint defects   237
        10.4.2 Surface defects   240
        10.4.3 Structural defects   241
    10.5 Maintenance faults and remedies   242
        10.5.1 Joint sealing   243
        10.5.2 Joint arris repairs   244
    10.6 Repair materials   246
    10.7 Overlaying and strengthening   246

11 Marine structures   248
    M.B. Leeming

    11.1 Introduction   248
    11.2 The need for repairs   249
        11.2.1 The underwater zone   249
        11.2.2 Above water   249
        11.2.3 The repairing of cracks   250
        11.2.4 The durability of repairs   250
    11.3 Repairs to deteriorated concrete   252
        11.3.1 Spalling and reinforcement corrosion   252
        11.3.2 Repairing cracks   253
        11.3.3 Repair of construction faults   253
    11.4 Repairs necessitated by damage to offshore structures   254
        11.4.1 Shipping damage   254
        11.4.2 Damage due to dropped objects   254
        11.4.3 Fire damage   254
        11.4.4 Gouges and abrasion   255
    11.5 Case studies of repair methods   255
        11.5.1 Shipping damage to the leg of an
                  offshore structure   255
        11.5.2 Punching shear failure to a caisson roof   255
        11.5.3 Gouges in concrete cell tops and caisson walls   258
        11.5.4 Damage to piles, jetties and other coastal
structures   258

x Contents

11.6 Repair materials   259
    11.6.1 Repair systems   259
    11.6.2 Testing repair materials   260
    11.6.3 Research into repair materials and methods for
        offshore structures   261
11.7 Repair strategies and conclusions   262

Index   **267**

Preface

As recently as 1976 textbooks on the design of reinforced concrete structures contained remarks such as 'compared with other building materials, reinforced concrete is distinguished for its very long service life. With proper service conditions reinforced concrete members can last an indefinite time without reduction of their load-carrying capacity. This arises from the fact that the strength of concrete does not decrease with time but, on the contrary, increases, and the steel embedded in it is protected against corrosion.'

Although today such statements appear somewhat naive, they are in fact essentially true. However, they need to be qualified, and as long ago as 1954 a survey showed that 'even on a moderately exposed site, the trouble-free life of reinforced concrete is likely to be short if the cover to the reinforcement is meagre, if binding wire projects towards the surface, or if for any reason the concrete is pervious to moisture'. In the rapid expansion of the construction industry during the 1960s such lessons were often forgotten and today in the UK we are seeing the result – a £500 million per annum programme of concrete repair.

In this book I have attempted to bring together experts from the field of concrete durability, repair and protection to provide a state-of-the-art report on current thinking, materials and techniques. The basic deterioration mechanisms and methods for the site investigation of distressed concrete structures are discussed first. Materials and techniques for repair are critically reviewed and relatively novel ideas for protection are discussed. The second part of the book consists of a series of case studies for various structure types. They are written by those who have immediate experience of both the technical and financial difficulties of concrete repair and maintenance programmes. Of necessity the information is presented for particular circumstances only and I cannot claim that the case studies are exhaustive. For example, low-rise housing has been consciously omitted, as has alkali–silica reaction where the solution is likely to be unique in every case.

I wish to thank all my contributory authors for their painstaking efforts amidst the pressures of modern life in the construction industry. Particular thanks, however, go to Jan Price who typed the complete manuscript for this book.

Geoff Mays

Biographic summary of contributors

Geoff Mays spent over 5 years on the design and supervision of construction of reinforced and prestressed concrete highway structures with Freeman Fox & Partners before joining Dundee University in 1977. There, as Associate Director of the Wolfson Bridge Research Unit, he was involved in extensive research into the structural use of adhesives in civil engineering. Since his appointment at the Royal Military College of Science (Cranfield) in 1984 he has conducted research for the Science and Engineering Research Council, the Department of the Environment and the Building Research Establishment on the structural consequences of patch repairs, ferrocement and the durability of lightweight concrete respectively. He is a member of British Standards Institution Committee CAB/17 dealing with the protection and repair of concrete structures. He has been Head of the Civil Engineering Group at the Royal Military College of Science (Cranfield) since 1988.

Tim Lees joined the Cement & Concrete Association in 1962 and left to join Taywood Engineering Ltd as the Manager of their Analytical Group in 1987. He has recently moved to W.R. Grace Ltd as Group Project Leader on admixtures and concrete technologies. He sits on a number of British Standards Committees and has been Chairman of Technical Committee CAB/6, Mortars, since 1987. He has lectured widely on the subjects of the chemistry of cement, acid and sulphate attack on concrete, and the analysis of hardened concrete and mortar. He has been involved in the Northwick Park long-term sulphate exposure trial since its inception.

Ken Berkeley joined Metal & Pipeline Endurance in 1954 and undertook design and installation work on marine and pipeline cathodic protection in the UK before studying cathodic protection practices in the USA. On his return to the UK he spent time with both Wallace & Tiernan Ltd and Sturtevent Engineering Ltd before accepting a directorship of Morgan Berkeley Ltd. In 1967 he established P.I. Corrosion Engineers Ltd and is now principal consultant of Corrosion Prevention Consultants. He was elected a Fellow of the Institution of Corrosion Technology in 1972 and a member of its Council in 1984. He has contributed over 20 papers on various aspects of cathodic protection and lectured throughout the world.

Jack Dixon has had many years experience dealing with research and consultancy in matters relating to the use of concrete.

He began his career dealing with research and development projects in the precast concrete industry before switching to British Railways and their track and structure development work. Became almost inevitably involved with durability problems and produced for BR's Civil Engineering Department probably the first ever *Guide to Concrete Repair* document.

A founder member and first Chairman of the Institute of Concrete Technology.

Tony Baker trained initially as a Geologist at Aston University before undertaking research at queen Mary College's Industrial Petrology Unit under Dr Alan Poole. He completed his PhD in 1978 on alkali-aggregate reactivity. He then spent six years at Taylor Woodrow Construction Research Laboratories where he became involved in concrete inspection and repair. In 1983 he moved to STATS Ltd as a project manager specialising in investigations of concrete structures. In 1985 he became a director of STATS and until recently was responsible for the engineering and materials divisions of STATS in St Albans.

Mike Leeming is a Chartered Civil engineer, has worked for local Authorities and Consultants on the design and construction of motorway and other major prestressed concrete bridges in the UK and abroad. He then joined CIRIA as Technical Manager of the Concrete in the Oceans research programme directing research into marine concrete for North Sea Oil structures. In 1984 he moved to Arup Research & Development to manage and carry out research and investigatory projects relating to the diagnosis and cure of deteriorated concrete, surface coatings on concrete, repairs and non-destructive testing. He is now with LG Mouchel & Partners.

Denis Northcott works in the Department of Transport, Highways Engineering Division and is responsible for pavement specifications, especially for concrete.

As a Royal Engineer he had experience of road and airfield construction in Malaya and in the Middle East, and railway construction and maintenance in Europe.

A period with contractors on a road scheme in Kent was followed by bridge design work with W.S. Atkins, before joining the Department of Transport. He was DTp member of the Advisory Panel for Concrete Road Construction for many years and co-author of the Manual for the Maintenance and Repair of Concrete Roads.

Jeff Keer is Training and Development Manager for the Balfour Beatty Group, which includes Balvac Whitley Moran, the structural repair company. He was previously Senior Lecturer in Civil Engineering at the niversity of Surrey, where he supervised a number of research projects in the field on durability and repair of concrete structures. He is presently a UK representative on one of the European committees harmonising standards in the repair industry.

Lawrie Tabor spent the earlier part of his career in the plastics industry, making synthetic resins, formulation and compounding thermoplastics and became chief chemist of a prominent moulding and extrusion company. He moved into the construction industry to develop a structural resin concrete for factory produced housing. After spending almost twenty years with Wimpey Laboratories in the development and testing of materials, latterly specialising in concrete repair materials and techniques, he is now with a major materials supplier.

He is an active member of the Concrete Society and the Plastics and Rubber Institute, serving on various construction and British Standards committes, has published a number of papers and lectures on concrete repairs.

PART ONE
Theories and Solutions

The behaviour of concrete   1

G.C. Mays
(Royal Military College of Science, Cranfield)

1.1 Historical development of concrete

The development of concrete as a construction material dates back several thousand years to the days of the ancient Egyptians, the Greeks and the Romans. These early concrete compositions were based on lime, although the Romans are known for their development of pozzolanic cement and lightweight concrete based on pumice. The domed roof of the Pantheon is testimony to the durability of this material. Apart from a brief revival during the Norman period when structures such as Reading Abbey and the foundations to Salisbury Cathedral were constructed from concrete, there was little further development until the eighteenth century. In 1756 Smeaton constructed the Eddystone lighthouse which stands to this day on Plymouth Hoe. In 1824 Aspdin took out his patent for 'a superior cement resembling portland stone' which was used for concrete in the construction of the Thames Tunnel. The Axmouth Bridge in Devon, a three-span all-concrete arch bridge, was built in 1887 and still carries traffic to this day.

The credit for the introduction of steel as reinforcement is variously attributed to Lambot in 1855 for ferrocement boats, to Monier in 1867 and to Hennebique in 1897 who built the first reinforced concrete frame building in Britain at Weaver's Mill, Swansea. It is interesting to note that one of Lambot's boats was still afloat 100 years later and reportedly in very good condition despite the seawater environment. However, the same cannot be said for Hennebique's first reinforced concrete bridge in Scotland which was reported in 1932 to be suffering from 'severe spalling as a result of the harsh environment'. Nevertheless, gunite repairs have extended its life to the present day.

Notable steps forward in this century have been the introduction of prestressed concrete by Freyssinet in the 1940s, the extensive use of reinforced concrete during World War II including the famous Mulberry Harbours, the rapid post-war concrete-building expansion prompted by shortages of steel, the motorway-building boom of the 1960s involving concrete pavements and bridges and most recently the contribution of structural concrete to very large offshore structures. Although the vast majority of concrete structures have performed satisfactorily for many years such progress has not been made without its problems. Structurally these range from the wind-induced failure of cooling towers at Ferrybridge Power Station to the progressive collapse due to a gas explosion of the Ronan Point tower block in East London in the 1960s. However, attention today is primarily focused on environmental attack which is substantially reducing the lives of many concrete structures around the world, in many cases due to corrosion of reinforcement steel.

1.2 Concrete — the material

The word 'concrete' comes from the Latin 'concretus' meaning compounded. Today concrete is understood to consist of a graded range of stone aggregate particles bound together by a hardened cement paste. Concrete is required to be strong, free from excessive volume changes and resistant to penetration by water. It may also need to resist chemical attack or possess a low thermal conductivity. Concrete's strength is derived from the hydration of the cement by water. The cement constituents progressively crystallize to form a gel or paste which surrounds the aggregate particles and binds them together to produce a conglomerate. Generally, the strength and permeability of the concrete are governed by its water–cement ratio. For high strength and low permeability the water–cement ratio should be low. Conversely, for ease of placing and compaction the easiest way of increasing workability is to increase the free water content, although nowadays chemical plasticizers are available to assist. In normal concrete the aggregate type has little effect on strength but it does affect workability. Normal procedures for the design of mix proportions involve selecting a water–cement ratio to provide the required strength and then determining water and additive content from considerations of workability and aggregate type. In high strength concrete the aggregate–cement ratio and aggregate type must also be taken into account. For all structural concrete it is also usual to specify a minimum cement content to ensure a level of alkalinity which inhibits any tendency for corrosion of embedded steel. Conversely, very high cement contents are inadvisable since early thermal contraction leading to cracking may result from the rise in temperature induced by cement hydration.

Concrete is a material which, although relatively strong in compression, is weak in tension and for structural members subject to tensile stress may be reinforced by steel bars. The effectiveness of reinforced concrete as a structural material depends on the following:

1. the interfacial bonding between steel and concrete which allows it to act as a composite material;
2. the passivating effect of the concrete environment to inhibit steel corrosion;
3. the similar coefficients of thermal movement of concrete and steel.

Alternatively, the concrete may be precompressed by applying load through tendons of high tensile steel anchored to the concrete. Under load the effect is to unload the compression and avoid significant tensile stresses.

1.3 Mechanisms and causes of failure

A discussion of the mechanisms and causes of failure of concrete is somewhat complicated by the fact that this material is often utilized as a composite in the form of reinforced concrete. Thus, consideration of failure mechanisms such as brittle or ductile fracture, fatigue, creep, instability or corrosion must take account of the nature and properties of all the materials which together constitute the composite.

Reinforced concrete is generally designed so that should failure occur it will be ductile in nature. Thus, in flexural members, the section properties are checked so that yielding of the reinforcement in tension precedes both the more brittle crushing of the concrete in compression and the very abrupt nature of diagonal tension shear failure.

Concrete itself is susceptible to cracking and eventual fatigue failure as a result of cyclic loading. However, for a given stress level the cycles to failure will increase as the concrete ages owing to the gain in static strength with time. If the concrete is reinforced or prestressed with steel then the fatigue properties of the steel tend to control structural performance. This is because tensile loads are carried by the steel and even quite extensive cracking in the concrete should not be too serious. Concrete also creeps with time under the action of sustained loads. The magnitude of the creep depends on material and environmental factors but most significantly on the level of stress and the age of the concrete when the stress is applied. In some circumstances, e.g. deflections of beams and losses in prestress, creep is detrimental; in others, for example, relief of stress concentrations, creep may be beneficial.

Leaving environmental attack and intrinsic changes within the concrete for the moment, perhaps the most likely causes of failure in concrete structures are due to accidental overload of instability. Hence civil engineers are required to ensure that the structures which they design are 'robust and stable'. By robust is implied the ability of a structure to avoid progressive collapse in the event of accidental overloading. Such overloading may be as a result of impact, e.g. ship collisions with bridge piers or more commonly these days high vehicles striking the soffit of highway bridges. Alternatively the overload may be due to the effects of flooding, e.g. scour around bridge piers, or to the effects of earthquake loads larger than those for which the structure was designed. Lastly, the unexpected effect may be due to blast loading, either accidental, such as gas explosions, or more deliberate, as a result of terrorism.

Concrete structures must also be designed to be both globally and locally stable. In other words there must be an adequate factor of safety against overturning under the most critical combination of horizontal and vertical loads. Also, individual members must not buckle under the action of compressive loads.

1.4. Intrinsic deterioration of concrete

It is generally accepted that reinforced concrete cracks in the hardened state under tensile structural loads. These cracks can be limited to acceptable levels by appropriate detailing. It is less well accepted that concrete cracks in both the plastic and hardened states as a result of the nature of the constituent materials. These non-structural cracks may be of three main types; plastic cracks which appear in the first few hours due to shrinkage or restrained settlement, early thermal contraction cracking which develops over the first few days following the exothermic peak, and longer term drying shrinkage.

These mechanisms are discussed in detail in Chapter 2 (section 2.2.1) but suffice it to say for the present that cracks in concrete may be significant in terms of the rate of subsequent reinforcement deterioration.

Problems have arisen with the use of high alumina cements for the construction of precast concrete products in the 1950s and 1960s. The material was first developed to resist sulphate attack, its resistance being derived from the absence of calcium hydroxide in its hydrated form. It also provides a very high rate of strength development which is of economic benefit in the precast industry. However, a number of dramatic failures in the early 1970s, notably in schools and swimming pools, led to a complete reappraisal of its use for structural concrete. It has since been found that the hydrates are unstable, particularly at higher temperatures and in the presence of water. Further details of the conversion process can be found in Chapter 2 (section 2.2.2). The practical interest in conversion lies in the resultant loss of strength to 20%–50% of initial. In view of the effects of conversion, high alumina cement is no longer permitted for structural concrete. Sulphate-resisting Portland cements are now available for use with concrete in contact with sulphate-bearing soils. Sulphate-resisting properties are also exhibited by concrete containing pulverized fuel ash (pfa) or ground granulated blastfurnace slag (ggbs) as partial cement replacements.

Another form of attack from within which is now seen on a significant scale is the phenomenon of alkali–aggregate or alkali–silica reaction. Under most conditions certain aggregates, for example siliceous cherts, appear susceptible to attack by the alkalis present within the surrounding cement paste. The aggregates gel and expand causing a disruptive effect on the concrete evidenced in the early stages by distinctive 'map cracking' at the surface.

Again, the process is discussed in detail in Chapter 2 (section 2.2.2). At present there is no known cure. The first demolition in Britain of a concrete structure weakened by alkali–aggregate reaction occurred over the M6 motorway in Warwickshire. The bridge, which was only 13 years old, showed particular deterioration in the cement-rich deck spans. There is also a growing concern over its possible effects in prestressed concrete. As the reaction is expansive, tendons can be further stressed thereby imposing additional overloading on anchorages which in the extreme case might burst.

1.5 Environmental attack

Environmental processes may cause salts, oxygen, moisture or carbon dioxide to penetrate the concrete cover and eventually lead to corrosion of embedded steel reinforcement. As steel corrodes it expands in volume causing cracking, rust staining and spalling of the concrete cover. The problem is widespread in all forms of concrete structures both in Britain and overseas. The two most significant penetration processes are the following.

1. Carbonation is caused by atmospheric carbon dioxide dissolving in concrete pore water and creating an acidic solution. This starts to neutralize the natural alkalinity of the concrete, and when the pH value around the steel bar falls below about 11.5 corrosion can begin.
2. Chloride salts may be presented in concrete either because calcium chloride was added at the time of construction as an accelerating admixture or because of ingress of de-icing salts, particularly in highway structures, ingress from seawater or spray in the case of marine structures or impurities in the aggregates and/or mixing water. Once chloride ions reach the reinforcing bar localized breakdown of the passivating environment can occur, forming sites of pitting corrosion attack. The subsequent rate of corrosion will be controlled by the level of moisture and oxygen present. In moist conditions corrosion cells are activated whilst oxygen fuels oxide formation on the steel reinforcement.

Finally, the simple process of freeze–thaw effects on concrete must be mentioned. Porous concrete which can become highly saturated with water is particularly liable to damage from frost attack. Use of de-icing salts containing chlorides greatly increases the chance of frost damage. For pavement-quality concrete air is entrained into the fresh concrete using an admixture which creates about 5% of small (about 0.5 mm diameter), stable, evenly dispersed air bubbles. These provide voids into which the ice can expand without disrupting the concrete. When the ice melts surface tension effects draw the water back out of the bubbles. To be effective the bubbles must not fill with water and air-entraining agents also coat the inside of the bubbles with a water repellent. However, with normal concretes, 5% air entrainment may result in a strength reduction of about 15%.

The deterioration of concrete due to these and other external agents, and the external effects leading to reinforcement corrosion, are covered in some depth in Chapter 2 (sections 2.2.3 and 2.2.4).

1.6 Prevention and repair

In the preceding section some comments have been made regarding techniques for minimizing the risk of failure due to both normal and abnormal structural loading. Cracking resulting from intrinsic effects within concrete is more likely to result in unserviceability, particularly in the case of water-retaining structures, than actual collapse. However, certain fundamental precautions both in the design process and during construction can minimize such effects.

It has already been mentioned that there is no known cure for alkali–silica reaction. Repair techniques are essentially those involving short-term control of the situation, e.g. reductions in moisture content and temporary support or propping. In the case of environmental attack it is the concrete cover to the reinforcement which traditionally has been the prime defence. Ideally this cover concrete should have low permeability to water, oxygen, chloride ions and water vapour. It is therefore unfortunate that the layer tends to be of poorer quality than the concrete in the heart of the member. Factors which may be beneficial in reducing penetration are the following:

1. increased depth of concrete cover;
2. low water–cement ratio to minimize capillaries formed as the fresh concrete 'bleeds';
3. high cement contents to provide a high level of alkalinity (note that this requirement may conflict with those for reducing early thermal contraction and with conditions for minimizing the risk of alkali–aggregate reaction);
4. efficient curing of adequate duration to minimize the formation of capillaries;
5. coatings or barrier treatments applied to the surface (see Chapter 5).

One repair method is complete reconstruction with members constructed from fresh concrete designed, placed and cured with the above factors in mind. However, it is often more economical to consider local patching of the damaged area. The patching process may be multi-staged and the choice of material for reinstating the cover is often a difficult one. The options are cementitious, polymer-modified cementitious or resin mortars generally selected in that order as the thickness of the patch decreases. The nature of these materials and the techniques for repair are fully debated in Chapter 4. Of prime importance to the long-term stress characteristics of the repair is the bond to the concrete substrate. There is also a need to consider the level of mismatch between the properties of the repair material and those of the original concrete with regard to resisting structural loading, thermal and creep effects and assessing the level of composite action between the two materials. An alternative method, particularly for protecting chloride-contaminated structures, is to use cathodic protection. A small current is applied to the reinforcement bars to render them passive again (see Chapter 5).

1.7 The extent of the problem

Concrete has been in service as a structural material for many centuries. In its massive form, when used for resisting essentially compressive loads, it is a reliable and durable material provided that precautions are taken to resist sulphate attack by ground water and freeze–thaw effects. The possibility of alkali–silica reaction must also be recognized.

Today, concrete structures need to be designed to be robust and stable. The number of actual structural collapses due to overloading is small although the possibility of earthquake damage or terrorist attack must not be forgotten. Similarly, the possibility of temporary overstress during construction must be considered.

It is unfortunate that concrete has to be reinforced with embedded metal when used to resist tension or flexural loading. It is the potential for corrosion of this embedded metal that is the prime cause of the majority of structural concrete deterioration that is now becoming evident. Indeed, it has been estimated that in 1988 the value of building and civil engineering repairs and maintenance carried out in the UK was of the order of £15 billion. Of this approximately £500 million per annum is spent on concrete repair. With such vast sums of money involved it is important that the construction industry recognizes the unwitting errors of the past and puts to good effect the lessons that have been learned for new construction. Buildings and other structures must be designed and costed with a planned maintenance programme in mind and when repairs become necessary their performance must be subsequently monitored.

Deterioration mechanisms 2

T.P. Lees
(Formerly Taywood Engineering)

2.1 Introduction

Cements are man-made chemicals which, when mixed with water, undergo a chemical reaction that is given the special name 'hydration'. The mass hydration products together with the unhydrated remnants of cement grains are called the 'cement paste' and the paste retains this name in dry, aged, hardened concrete. The aggregate particles are dispersed through and held together by the matrix of hardened paste (Fig. 2.1).

To a very large extent the chemistry and particularly the physical structure of the cement paste determines the durability of concrete and the nature

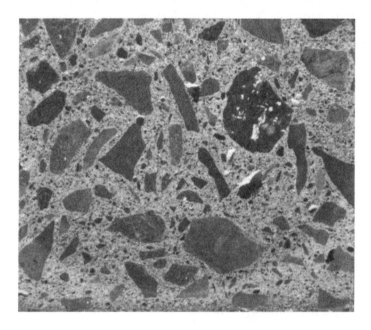

Fig. 2.1 Polished surface of concrete.

of any deterioration (Lea, 1970). Although detailed discussion of the chemistry of cement is outside the scope of this book, some chemistry and physics must be invoked since deterioration of concrete (or mortar) arises because of chemical or physical changes.

2.1.1 Cement

Portland cements are composed primarily of calcium silicates which constitute about 80% of the mass. The remaining 20%, comprising aluminates, ferrites and other compounds, although contributing little to cementing properties, is essential in enabling cement to be manufactured economically. Thus the aluminium and iron naturally present in the raw materials permit the formation of the silicates in the cement kiln at a much lower temperature than would be the case if they were absent. Because Portland cements are made using locally available raw materials, their composition varies from source to source. Figure 2.2 is a magnified section of a grain of Portland cement and clearly shows the major compounds.

Materials used as alternative cementing agents in conjunction with Portland cement, such as pulverized fuel ash (pfa) and ground granulated blast furnace slag (ggbs), have broadly similar compositions or at least give similar hydration products to Portland cement. The only non-Portland cement used in quantity is high alumina cement (HAC) which is composed primarily of

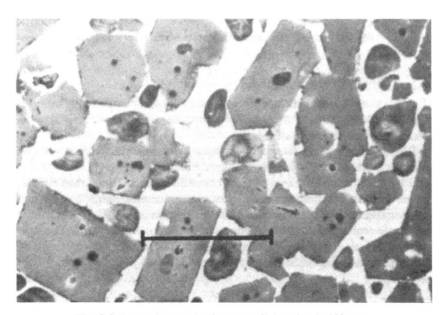

Fig. 2.2 Photomicrograph of cement clinker (bar is 100 μm).

calcium aluminates and contains only small amounts of calcium silicates. Figure 2.3 is a triangular diagram which demonstrates that all the cementitious materials commonly used in concrete or mortar are broadly related.

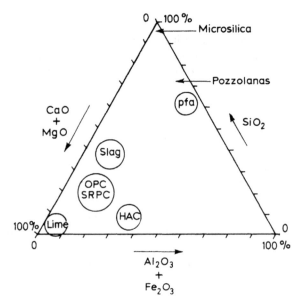

Fig. 2.3 Triangular diagram of common cementitous materials.

2.1.2 Hydration of cement

The reaction of water with the cement compounds is exceedingly complex and only those details of hydration which play a major role in potential deterioration will be considered.

The main cementing reaction of Portland cements is the hydration of the calcium silicates to form calcium silicate hydrate gel, C–S–H*. It is called a gel because it has no clearly defined crystalline structure and the formula given recognizes that its actual composition is indeterminate. The reaction is accompanied by the formation of calcium hydroxide, $Ca(OH)_2$, which eventually forms about 20% of the volume of the cement paste. The two calcium silicates present in Portland cement, C_3S and C_2S, give essentially the same product but C_3S reacts faster than C_2S.

---

*In cement chemistry a shorthand notation is used to represent complex compounds, e.g. C, calcium oxide; S, silica; H, hydrate; A, alumina.

The other hydration products whose chemistry influences possible deterioration are formed from one of the lesser cement compounds, tricalcium aluminate, C_3A. The hydrates of this compound have a role in sulphate attack and also in the ingress of free chloride ions which affects the corrosion of embedded metal.

Alternative cementing agents used with Portland cements react with its hydration products in a very complex way but generally to form C–S–H. HAC forms calcium aluminate hydrate but little or no calcium hydroxide.

All the above reactions take time to complete and hydration, or at least molecular rearrangements, can still be occurring after months or even years. In addition, reaction is accompanied by some heat generation, the amount depending on the chemistry of the cement.

2.1.3 Physical structure of cement paste

Because we are dealing with a chemical reaction, the quantity of water required for complete hydration is fixed by the chemistry of the cement. For Portland cements it can be shown that, after about one month's hydration under normal conditions, 100 g of anhydrous cement will have reacted with about 23 g of water. This can be expressed as a water–cement ratio W/C of 0.23. However, to achieve full hydration, an excess of water is required, and extra water is also needed to give the paste sufficient fluidity to render the concrete workable.

With well-graded aggregates, the free water content of fresh concrete is in the range 150–200 litres for each cubic metre of concrete, depending on the required workability. Water contents much outside this range are not practical unless admixtures are used or the aggregate grading is adjusted. For example, with a superplasticizer a free water content of about 120 litres/m^3 may be possible, and with the fine sands used in mortars the water content may be as high as 350 litres/m^3

The water content is independent of the cement content; thus normal concrete with a cement content of 250 kg/m^3 will have a free W/C in the range 0.6–0.8 and for concrete with a very high cement content of 500 kg/m^3 the free W/C will be in the range 0.3–0.4. Even in the latter case, W/C is higher than that needed for hydration (0.23) and this extra water will in due course dry out if climatic conditions or the location of the concrete permit. The volume formerly occupied by this extra water exists as a system of pores communicating to the surface of the concrete. This property is known as the capillary porosity and the volume is obviously greater the higher the original water–cement ratio. Incomplete hydration, usually caused by lack of curing, further increases the volume of capillary pores.

The capillary porosity in the paste makes concrete permeable to liquids and gases as demonstrated in Fig. 2.4. The capillary porosity is not the only route by which gases and liquids can permeate through concrete because

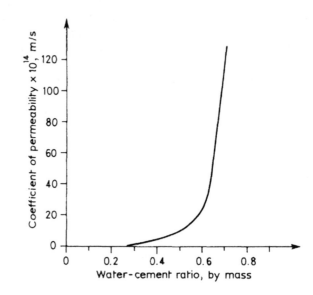

Fig. 2.4 Typical relationship between permeability and water–cement ratio in concrete.

the hydrated gel also has a pore system, although the gel pores are considerably finer than the capillary pores. A third pore system is provided by air voids resulting from incomplete compaction and these voids are much larger than the capilllary pores. In air-entrained concrete, many small discrete air bubbles are formed in the paste connected to one another by the gel pore system.

The above account of the structure of the cement paste is deliberately simplistic. In practice, the volume of hydrated gel is significantly larger than the volume of its constituent parts (cement and water) so that it can eventually fill part of the capillary pore system. With low water–cement ratios, the capillary pores become discontinuous so that the permeability is significantly reduced. However, because of the gel porosity ordinary concrete can never be completely impermeable.

2.1.4 Simplified model of concrete

Figure 2.1 shows the individual particles of a graded aggregate dispersed through and embedded in a hardened cement paste matrix. The matrix consists of a continuous gel of C–S–H which contains crystals of calcium hydroxide and other hydrates of the aluminates and ferrites. Dispersed through the gel are the unhydrated remnants of cement grains which were originally too large to hydrate completely.

The gel has an inherent porosity with very fine gel pores (about 4 nm* in diameter) and considerably larger capillary pores whose volume and pore diameter increase with higher original water–cement ratio (the pore diameter is about 200–500 nm for $W/C = 0.65$). In addition, the paste will contain air voids resulting from incomplete compaction. In all except oven-dry concrete, part or all of the pore system is filled with 'pore fluid' which is essentially a solution of NaOH or KOH saturated with $Ca(OH)_2$. In fresh green concrete the paste will be warm or even, in well-insulated high cement content concrete, hot.

2.2 Deterioration of concrete

Deterioration of concrete can take many forms, real and apparent. To the general public, the most obvious is the change in appearance caused by natural weathering. Many surveys have shown that corrosion of reinforcement is the single most common source of damage and it is usually clear that inadequate concrete, inadequate cover to the reinforcement or the presence of impurities (and sometimes all three) is the prime cause. Much of the deterioration of concrete can only occur in the presence of water since aggressive agents will penetrate concrete and react harmfully with the cement paste only when dissolved in water.

It should be stressed that much the greater part of the volume of concrete produced does not suffer any deterioration. Good quality concrete (i.e. with adequate cement content, low water–cement ratio, well compacted and cured) is relatively immune to harmful effects and is likely to be quite unaffected if water is absent. Nevertheless, a number of processes whereby concrete can become damaged exist, and these will be considered in turn.

2.2.1 Early deterioration

Damage caused at an early age is perhaps not strictly 'deterioration', but it is relevant since cracks can provide routes for the ingress of harmful compounds, particularly those promoting corrosion of embedded metals.

Plastic concrete deterioration

Fresh concrete is a suspension of aggregate particles in a fluid paste of cement and water. The paste has a lower density than the particle density of normal aggregates so that gravity will tend to pull the heavier particles downwards.

---

1 nm = 1 nanometre = 10^{-9}m.

This is called 'sedimentation'. For example, a paste with $W/C = 0.6$ will have a density of about 1750 kg/m³ whereas a typical aggregate will have a particle density of 2650 kg/m³. As the particles descend, the paste is displaced upwards and there can be a further separation of the heavy cement particles sedimenting within the paste so that water is displaced upwards. This is called 'bleeding'.

In very stiff cohesive concrete, sedimentation and bleeding will be minimal, whereas in high water content concrete the effect may be considerable. In some cases, a layer of clear water will be seen on the top surface of concrete.

It should be noted that high slump concretes produced by the use of admixtures will not show untoward bleeding because the extra workability has been achieved by better dispersal of the cement within the paste and this will have the effect of making the paste more cohesive. It should also be noted that bleeding in itself is not necessarily deleterious – rather the reverse, since the effective water–cement ratio of the sedimented concrete will be lower.

The mechanisms whereby sedimentation and bleeding can be deleterious are fivefold.

1. The aggregate in the concrete underneath horizontal reinforcement sediments downwards and the rising bleed water becomes trapped under the bar. The final result is that the underside of horizontal steel is not in good contact with the hardened paste and in some cases there may be a lens-shaped void under the bar (Fig. 2.5). It is often observed that reinforcement in this state has corroded on its bottom surface.

2. Rising bleed water becomes trapped under large aggregate particles. In some working operations flat particles tend to become aligned in a horizontal plane and voids under such a particle near the surface of concrete, when water filled, can lead to frost damage such as pop-out of the particle.

3. Sedimentation and bleeding lead to stiffening of the concrete, since water is being lost, and if an obstacle prevents further localized sedimentation the concrete can crack. This is usually seen only in deep sections because the amount of sedimentation is greater.

   The most common 'obstacle' is horizontal steel at the top of the section. Figure 2.6 is a schematic representation of the effect which is called 'plastic settlement'. When it occurs over reinforcement, the cracks may eventually lead to corrosion of the steel.

   Plastic settlement cracks can also occur in sections where there is a significant change in the depth of the concrete, e.g. in trough and waffle slabs.

4. When the rate of evaporation of bleed water is faster than the bleeding rate, an effect called 'plastic shrinkage' can occur. The removal of water

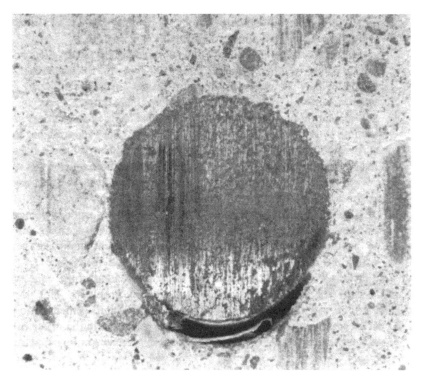

Fig. 2.5 Void under reinforcing bar due to plastic settlement of fresh concrete.

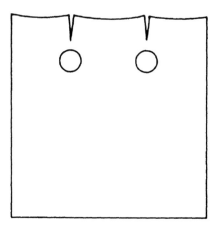

Fig. 2.6 Plastic settlement over reinforcing.

by evaporation or suction leads to a reduction in the bulk volume of the concrete and if the concrete cannot follow this change in volume fast enough, e.g. by settlement, then cracks will result.

Plastic shrinkage cracking is most common on flat slabs placed during hot or windy weather which are not protected almost immediately after laying. A particularly severe form of plastic shrinkage cracking occurs when concrete is placed on a high-suction substrate.

5. Sedimentation of coarser aggregate particles leads to an increase in paste volume at the top of a pour. If bleed water does not evaporate but is reabsorbed into the top surface, the top of the section will have a higher effective water–cement ratio than the bulk of the concrete. For this reason, coring for strength determination or chemical analysis is not carried out in the top 10% of a lift (Concrete Society, 1987).

With floor slabs, particularly those worked to produce smooth flat finishes, excessive bleeding and too early trowelling exacerbate this effect so that the top surface is a layer of high water–cement ratio paste (laitance) which has much reduced resistance to wear and abrasion.

In all cases of plastic deterioration, an improvement in the cohesiveness of the concrete will reduce or even eliminate any problems. This improvement is generally achieved by a reduction in the water–cement ratio either by an increase in the cement content or by a reduction in the water content, e.g. by using admixtures. Attention to the aggregate grading may also be necessary.

Settlement effects can be eliminated by judicious revibration an hour or so after completion of the initial compaction and plastic shrinkage can be reduced or eliminated by early protection of the concrete from evaporation.

Thermal effects

The hydration of cement is exothermic, i.e. heat is evolved. Measuring the change in the rate of heat evolution with time is a powerful experimental technique for studying cement hydration mechanisms.

A typical output from such an experiment is shown in Fig. 2.7. There is an initial heat output as soon as the water is added which comes from reaction of free lime, the so-called 'heat of wetting'. When this surge is complete there is a 'dormant period' and then, corresponding broadly to final set and the beginning of strength development, a broad peak is seen. This is caused by the hydration of C_3S and C_3A. This peak starts about one to three hours after the addition of water and has reached a maximum after about a day's reaction. The peak reduces but does not reach zero since C_2S is still reacting and giving out heat after seven or more days.

The rate of heat evolution is measured in watts and the area under the curve, the heat of hydration, in joules. The heat of hydration of ordinary Portland

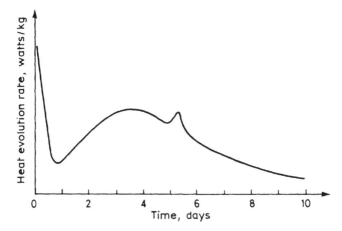

Fig. 2.7 Heat evolution of hydrating cement.

cement (OPC) is about 250 kj/kg. The temperature rise of insulated concrete is given very approximately in degrees Celsius by

$$\frac{\text{cement content } (kg/m^3) \times \text{cement heat of hydration } (kJ/kg)}{\text{mass of } 1m^3 \text{ concrete } (kg/m^3) \times \text{specific heat of concrete } (kJ/kg°C)}$$

The specific heat of typical conventional concrete is about 1 kJ/kg° C so that we can expect a 10° C rise in temperature for each 100 kg of cement in 1 m³. In concrete which is not insulated the heat generated is dissipated by cooling to the surroundings and slow heat evolution will cause a lower temperature rise than rapid heat output.

The thermal expansion of concrete depends on its mix proportions and on the type of aggregate present but values for the linear coefficient of thermal expansion can be assumed to be in the range $(6-13) \times 10^{-6}/°C$. For a concrete containing 350 kg cement/m³, with a gravel aggregate, the thermal expansion could be as high as $35 \times 13 \times 10^{-6} = 450$ microstrain. If the concrete can expand on heating and contract on cooling without any restraint, cracking should not occur.

One of the most common mechanisms whereby thermal movement causes cracking is when walls are cast on to hardened bases. When the new concrete cools and contracts, the junction between the wall and the base restrains the contraction at that level.

Another cause of thermal cracking is with massive concrete where outer layers of cooling concrete are counteracted by a stable or even expanding core of still warm concrete. In this case, insulation to inhibit the cooling of the layers, so that all the concrete is at a relatively uniform temperature, should prevent differential movements.

The magnitude of temperature effects can be reduced by using cements with low heats of hydration such as blastfurnace slag or pozzolanic cements. Portland cements with low C_3S and high C_2S contents, i.e. the reverse of Opc, are classified as low-heat Portland cements but are infrequently manufactured at present.

Drying shrinkage

As discussed earlier, the free water of fresh concrete evaporates on drying to leave a system of capillary pores. With properly cured concrete this loss of capillary water leads to no (or at least insignificant) volume change. However, during atmospheric drying, water that is not combined chemically as cement hydrates remains in the gel pore system. When this water evaporates there is a shrinkage of the cement gel although the magnitude of the shrinkage is significantly less than would be calculated merely from a consideration of the volume of water loss, probably because some of the loss is accommodated by an increase in pore volume.

The shrinkage of the gel fraction is restrained by the presence of aggregate. In concrete the volume of aggregate has a major influence on the shrinkage not only because of the restraint it offers but also because a corollary of more aggregate is less paste. A high original water–cement ratio leads to an extended paste system which can therefore shrink more. Thus, although shrinkage of normal concrete is inevitable, the magnitude of the shrinkage is reduced for low cement contents and particularly low water–cement ratios.

Shrinkage is properly accommodated by provision of contraction joints and by normal design procedures. However, in some instances widespread cracking can occur if the outer layers of the concrete dry and shrink much more rapidly than the interior. It is likely that nearly all concretes have an extensive system of microcracking induced by drying shrinkage and also by thermal movements. It is only when the cracks become macro-sized that problems may occur.

A special problem is posed by the use of shrinkable aggregates. There is no clear-cut recognition of the causes of aggregate moisture movement since it has been recorded that, even within a single quarry, rocks from different positions have widely differing shrinkages. Whereas the shrinkage of normal concrete is of the order of 0.03%, that of concretes made with shrinkable aggregates can be over 0.1%.

In the UK, the largest group of aggregates exhibiting damaging shrinkage are dolerites, although many doleritic aggregates are completely satisfactory. Severe damage is usually limited to concretes of thin section. Damage, when it does occur, is usually observed at 6–12 months, i.e. when the weather, particularly frost, has accentuated the cracking. Building Research Establishment (BRE) *Digest 35* gives recommendations for using such aggregates (BRE, 1968).

2.2.2 Deterioration arising from concrete ingredients

Mention has been made of shrinkable aggregates in the previous section. One of the most severe causes of deterioration is the presence of undesired impurities. For example, the presence of excess sulphate in freshly made concrete, most commonly from contaminated aggregates, will cause severe sulphate attack. Cement that does not comply with specification limits for sulphate is an unlikely but not completely unknown source. Regrettably the addition of gypsum plaster to concrete by unsupervised operatives attempting to decrease setting times has occurred more often than it should. However, the mechanisms of sulphate attack from these causes are similar to, though usually more severe than, those arising from sulphates entering the hardened concrete from external sources and will therefore be considered in section 2.2.3.

Similarly, corrosion of reinforcement induced by chlorides follows similar paths regardless of whether the chloride was incorporated in the fresh concrete as an admixture or impure aggregate or has ingressed into the hardened concrete at a later date. The effect of chlorides on the durability of reinforced concrete is discussed in detail in section 2.2.4.

Conversion of high-alumina cement

The hydration of HAC gives a gel with similar physical properties to that given by Portland cement but with a different chemistry. The HAC gel is composed primarily of calcium aluminate hydrates with idealized formulae CAH_{10} and C_2AH_8. Both these hydrates will gradually rearrange to form the most stable calcium aluminate hydrate C_3AH_6. However, whereas in Portland cement the early aluminate hydrates rearrange to C_3AH_6 without significant change in volume, the change in HAC is accompanied by a considerable reduction in solid volume, e.g. approximately 50% reduction in going from CAH_{10} to C_3AH_6.

The rearrangement is extremely slow at temperatures below about 18 °C and does not occur to any great extent in completely dry conditions. However, in wet conditions and at temperatures of 25–30 °C the change can be rapid (Neville, 1977). The consequence of the reduction in solid volume of the hydrates is an increase in the porosity of the cement hydrate paste with a corresponding decrease in the strength of the paste and hence the concrete.

W/C for complete hydration of HAC is about 0.5 so that the recommended free *W/C* of 0.35 for HAC concrete leads to a large reservoir of unhydrated cement. As a consequence, the conversion of the hydrates, for which water is needed, permits access of water to the unhydrated cement grains and their hydration gives new products which fill the induced porosity.

The consequences of HAC conversion when the above recommendation is not followed and too much water is used are a lowering of strength arising from an increase in porosity which, in turn, can lead to distress in the concrete which can no longer accommodate the loads applied. Thus cracking of HAC concrete members is not caused directly by conversion but by the weakening it induces.

The greatest loss of strength occurs with concrete of high water–cement ratio which has been allowed to get warm during early hydration and is subsequently exposed to a warm humid environment. There is an insignificant loss of strength in concrete of low water–cement ratio which has not heated during hydration, either because of use in small sections or because cooling has been applied, and which is exposed to dry and/or cool surroundings.

Alkali–silica reaction

The terminology causes confusion because an alkali is normally the opposite of an acid and the strongest alkalis are hydroxides which are present in considerable amounts in hydrated cements. However, in the early days of chemistry trivial names were widely used and a certain group of elements were called the 'alkali metals', the two most abundant ones being sodium and potassium. It is the reaction of these two elements, albeit as their hydroxides, which is referred to in the alkali–silica reaction (ASR). Other 'alkali' metals, more properly group I elements such as lithium and rubidium, may well take part in the reaction but they do not occur in concrete-making materials to any significant extent.

Potassium and sodium, however, occur widely in natural materials and in cement are present mostly in soluble forms. The pore water in concrete contains quite high levels of sodium and potassium hydroxides which help to give concrete its high alkalinity (using the word in its proper meaning).

In many industries, including those dealing with concrete materials, it is conventional to express elements as their oxides although the oxides do not necessarily exist as such. Thus sodium and potassium are referred to in terms of Na_2O and K_2O. When dealing with ASR it is convenient to bring the two together as 'equivalent Na_2O'. To do this quantitatively, the amount of K_2O is divided by the molecular weight of K_2O, and multiplied by the molecular weight of Na_2O i.e. 61.98/94.20 (= 0.658), and is then added to the amount of Na_2O. To simplify discussion of the reaction in what follows, reference to Na_2O (sodium hydroxide) should be taken to imply K_2O (potassium hydroxide) as well.

The etching of glass by strong hydroxide solution demonstrates clearly that reaction is taking place between the silica in the glass and the hydroxyl ions. In concrete containing siliceous aggregates, e.g. quartz in the form of flints, the surface area of the silica and the concentration of sodium hydroxide in the pore fluid are not normally sufficient to give any significant reaction, although chemical analysis of concretes more than about five years old often shows that some of the silica has reacted.

However, certain types of siliceous materials which may occur in deposits used as concreting aggregates have a very fine grain structure which allows pore water to enter the particles and presents a large overall surface area (including the internal surfaces) to it. Also, cements contain some sodium, the amount depending on the nature of the raw materials used to make them. If such a fine-grained siliceous aggregate is used with a cement whose sodium content, together with the amount of cement present, is sufficient to give a critical sodium hydroxide concentration in the pore liquid then a significant amount of reaction can take place.

In the deleterious form of the reaction, the reaction product is a sodium-rich silicate gel which has the property of taking up large amounts of water with a consequent increase in gel volume. In some cases the expanding gel fills pores and voids in neighbouring aggregate particles and cement paste but in other cases the expanding gel exerts so much pressure that cracking occurs. If the concrete dries the gel shrinks, opening up the cracks still wider.

It appears that the nature of the gels which cause deleterious expansion is dependent on the sodium–silica ratio. Therefore if the reactive particles in the aggregate are more than a certain level, or if they are so fine that sufficient reaction can take place, the gel formed will have a sodium–silica ratio which does not imbibe water as aggressively and therefore does not expand.

A number of factors are necessary for the reaction to be deleterious. The soluble sodium content of the concrete must be high enough to give the pore liquid a critical concentration of sodium hydroxide. (Note that soluble sodium compounds from any source can end up as sodium hydroxide by exchange with calcium hyroxide.) There must be sufficient reactive particles present to give enough reaction product for the total swelling to disrupt the concrete. However, the amount of particles reacting must not be so great as to give a non-swelling gel. There must be a source of water for the gel to imbibe. Lastly the concrete must not be sufficiently porous to accommodate the gel expansion.

The pattern of cracking associated with ASR is caused by expansion of limited numbers of individual particles distributed at random through the concrete (Fig.2.8). Reactive particles on or near the surface of concrete do not expand and crack because any gel produced is easily accommodated by exudation to and away from the surface.

2.2.3 Concrete deterioration caused by external agents

As discussed earlier, concrete is inherently porous although with 'good'

Fig. 2.8 Unrestrained map cracking on parapet wall due to alkali-silica reaction.

concrete the permeability to gases and liquids is very low. However, the outer 'skin' of concrete cast against shuttering or trowelled in some way is cement and water rich and is more porous than the body of concrete. The skin may be quite thin – less than 1 mm (Fig. 2.9).

Agents which are aggressive to cements, e.g. acids and sulphates, can only penetrate concrete when dissolved in water and even aggressive gases only react with the cement hydrates in the presence of water. Two agents associated with the deterioration of concrete are carbon dioxide and chlorides. They do not attack concrete as such (although very high concentrations of chloride may do so) but they promote the corrosion of embedded metals and will be dealt with in section 2.2.4.

It should be pointed out that carbon dixoide (CO_2), which causes carbonation, is in practice positively beneficial to concrete since carbonated concrete is stronger and more resistant to penetration by aggressive agents. The confusion arises because excessive carbonation is associated with poor concrete and the benefits of carbonation are not sufficient to turn it into good concrete.

Weathering

Weathering is the deterioration of the outer skin of concrete caused by the

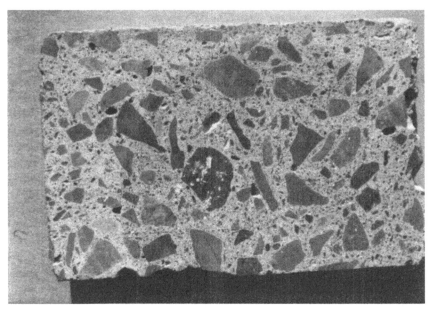

Fig. 2.9 Section of concrete cement-rich (upper) surface.

effects of rain, frost, sunlight and atmospheric pollution. The result is a change in appearance, usually considered to be for the worse with concrete and modern materials but sometimes acclaimed as the 'noble patina of age' with 'traditional' materials (Hawes, 1986).

Fine particles of atmospheric dirt settle on to building surfaces and rain, also contaminated with dirt, falls on to and runs off the same surfaces. When the outer skin is porous it acts as a filter so that the dirt particles are trapped in the pores. Rapid flow of water across the surfaces may wash the particles out but usually only to redeposit them when the direction or rate of water flow changes.

Some of the particles are biological in origin and algae or lichen growth develops from them, which is often the cause of discoloured concrete surfaces.

In built-up areas in particular, the atmosphere is polluted with sulphur dioxide gas (SO_2) which appears to accumulate on surfaces. It is acidic and reacts with alkaline concrete. Under the influence of solar radiation, catalysed by other pollutants, it forms calcium sulphate or gypsum. Dissolution and recrystallization of this in the pores of the surface leads to disruption and 'fretting' of the surface. This mechanism leads only to removal of the outer skin of concrete and the underlying body of the concrete is unaffected. With some porous natural stones, however, the effect can be deeper and more severe. Because the outer skin is porous and directly exposed to the atmosphere, those surfaces exposed to the rain are easily saturated. Subsequent freezing can cause fretting of the surface.

Calcium hydroxide, a major component of hydrated cement, is soluble in water to a limited extent (about 1 g litre). As young concrete dries out, dissolved calcium hydroxide moves towards the surface where it reacts with atmospheric carbon dioxide to form almost insoluble calcium carbonate. Usually this reaction takes place **in** the surface layer but in some circumstances it takes place **on** it to give lime bloom. This is similar to efflorescence and is often given that name. If water flows through very porous concrete or through poorly detailed joints, the calcium hydroxide can be carried to the surfaces in large quantities and carbonation then gives very unsightly growths and stalactites. Although these can be quite extensive, it is rare for sufficient calcium hydroxide to be removed to change the properties of the main concrete significantly.

Acid attack

Hydrated cement is alkaline and reacts with acids (Harrison, 1987). The effect is generally to dissolve the cement hydrates as calcium salts. For example, hydrochloric acid (HC1) is used in the chemical analysis of hardened concrete to dissolve the cement completely, the products being soluble calcium chloride, iron and aluminium chlorides and silicic acid. All cements except HAC are equally susceptible. However, there are a number of mitigating circumstances and in practice acid attack is only a problem in industrial processes, in sewers and in circumstances where the concrete is exposed to rapid flows and considerable volumes of acid.

Concrete made with insoluble aggregates, i.e. most of those used other than limestones, exposes progressively less cement hydrate to the acid as acid attack proceeds and the aggregate is exposed. Also the silica in the cement is dissolved as silicic acid but this readily polymerizes to form silica gel which is insoluble and 'clogs up' the pores in the hydrated cement, thus inhibiting further ingress of acid.

The aggressiveness of the acid depends on the solubiility of its calcium salt. Thus hydrochloric acid dissolves the cement to form calcium chloride which is very soluble and is consequently very aggressive. In contrast, a natural acid present in acid soils, humic acid, forms calcium humate which has a low solubility and hence the acid is not very aggressive. Oxalic acid forms insoluble calcium oxalate and is totally non-aggressive. The surfaces of the paste exposed to the acid becomes coated with the insoluble salt so that reaction ceases.

Sulphuric acid forms calcium sulphate which has a limited solubility but when the water containing sulphuric acid has a vigorous flow, as may occur in sewers, the calcium sulphate can be physically removed by scouring as well as dissolution and considerable attack can take place. Sulphuric acid can be formed in sewers by biological processes acting on neutral sulphates. Sulphate attack (see next) does not appear to occur in acid conditions.

Because the acid is neutralized when it reacts the effect is usually limited to the surface but if sufficient acid is present the reaction is progressive so that in extreme cases the concrete is eventually completely destroyed. However, the calcium salts formed may be absorbed into the concrete and give rise to other deterioration mechanisms. For example, hydrochloric acid may leave high concentrations of calcium chloride in the unattacked concrete which can promote corrosion of the reinforcement.

A special case of acid attack can be caused by dissolved carbon dioxide in soft water which behaves as a solution of 'carbonic acid' H_2CO_2. This forms not insoluble calcium carbonate but relatively soluble calcium bicarbonate. Attack is only significant with high flows of soft water saturated with carbon dioxide and is generally a surface effect since precipitation of silica gel as a result of cement hydrate dissolution inhibits further attack. Indeed cases have been observed where such waters acting on limestone aggregate concrete have dissolved the aggregate without significant attack on the cement paste.

When a finite quantity of acid is involved the use of limestone aggregate is beneficial since it is itself dissolved by acids and gives extra neutralizing power. There is also a more uniform loss of surface compared with the rough exposed aggregate surface left with insoluble aggregates. However, with continuous flows, attack is more pronounced.

Sulphate attack

Two of the components of cement paste, calcium hydroxide and calcium aluminate hydrates, react with sulphate ions in solution. The solid reaction products have larger volumes than the initial compounds so that disruption of the paste occurs.

The reaction of calcium hydroxide with dissolved sulphate ions depends on the nature of the sulphate. Chemical reactions are written as equations with the left-hand side of the equation being the reactants and the right-hand side the products. The two sides are connected by an 'arrow' denoting the direction of change. However, most ionic reactions do not always go to completion but achieve an 'equilibrium state' and can then proceed in either direction. This is exemplified by the reaction between calcium hydroxide and sodium sulphate:

$$Ca(OH)_2 + Na_2SO_4 \rightleftharpoons CaSO_4 + 2NaOH \tag{2.1}$$

The reaction goes from left to right until the concentration of sodium hydroxide (NaOH) builds up sufficiently to stop the reaction. If further NaOH is formed the reaction can then go from right to left. Thus the efflorescence on some clay bricks is caused by the reaction of NaOH from cement hydration reacting with insoluble gypsum ($CaSO_4$) in the brick to form soluble sodium sulphate which can then migrate to the brick surface and, on drying, crystallize. Equation (2.1) denotes the 'equilibrium' condition; for example, if the concentration of sodium sulphate is 2% only about one-fifth of the sulphate forms calcium sulphate (Lea, 1970).

In contrast, the equation with magnesium sulphate

$$Ca(OH)_2 + MgSO_4 \rightarrow CaSO_4 + Mg(OH)_2$$

goes wholly from left to right, because magnesium hydroxide is very insoluble and therefore the reaction from right to left is insignificantly slow. As a result magnesium sulphate solutions react completely with the calcium hydroxide.

When the reaction goes to completion, as with magnesium sulphate, all the free calcium hydroxide may be changed to calcium sulphate. Calcium silicate hydrate gel (C–S–H) is only stable in the presence of calcium hydroxide and so it partially decomposes to give more calcium hydroxide and hydrated silica. Thus with strong solutions of magnesium sulphate not only is there a disruption caused by the physical forces of expansion but there can also be a progressive breakdown of the silicate structure.

Ammonium sulphate, usually only encountered in industrial or agricultural environments, is also particularly aggressive because its reaction with calcium hydroxide goes to completion. The ammonium hydroxide formed is continuously evolved as ammonia gas and water.

It is clear that calcium sulphate cannot react with calcium hydroxide in a significant manner because both sides of the equation will be the same.

This reaction with the calcium hydroxide is an important part of sulphate attack and is the primary reason why magnesium sulphate (and ammonium sulphate) is so much more aggressive than sodium sulphate and calcium sulphate. The theoretical expansion in changing from solid $Ca(OH)_2$ to solid $Ca(SO_4)$ is about 2.2 times.

Carbonation of hydrated cement will be discussed more fully in section 2.2.4 but the effect is to convert calcium hydroxide to insoluble calcium carbonate, with very little volume change. Calcium carbonate does not participate in reactions with neutral sulphate solutions and thus carbonated concrete is not susceptible to this part of sulphate attack. The better resistance of 'real' concrete to sulphates compared with that expected from consideration of laboratory experiments is probably due to the existence of a carbonated layer at the surface.

It is also likely that the presence of dissolved carbonates in seawater makes it less aggressive than its sulphate content would suggest. In a competition for reaction with calcium hydroxide between sulphate and carbonate, the latter will win because of the insolubility of calcium carbonate.

The other major reaction involving sulphates is with the calcium aluminate hydrates. Controlled amounts of calcium sulphate are added to Portland cements to control their setting and early behaviour by moderating the rate of reaction of the C_3A with water.

Two forms of calcium sulphoaluminate hydrate are formed: a 'low' form, monosulphate, $C_3A.CaSO_4.12H_2O$; and a 'high' form, ettringite, $C_3A.3CaSO_4.31H_2O$. During the hydration of Portland cement, the sulphate retarder is consumed by the C_3A to form ettringite which subsequently rearranges to monosulphate. When all the sulphate has reacted, the remaining C_3A then hydrates. However, at this stage it is being hydrated by super-saturated calcium hydroxide solution and the hydrate formed is C_4AH_{13} or C_4AH_{19}. There is some controversy as to which is actually present. There is probably a slow change under normal conditions to form the stable hydrate C_3AH_6 and this compound predominates in autoclaved concrete.

Therefore, when sulphate ions in aqueous solution react with hardened concrete we are considering reactions typified by

$$C_4AH_{13} + 3CaSO_4 \rightarrow C_3A.3CaSO_4.31H_2O + Ca(OH)_2$$

The reactant is calcium sulphate. When other sulphates are involved their reaction with the calcium hydroxide is the source of calcium sulphate. The theoretical expansions (or even contractions) for the possible reactions are given in Table 2.1. C_3AH_6 reacts considerably more slowly than the other aluminate hydrates and this explains why HAC and autoclaved Opc concrete are very resistant to sulphate attack.

Table 2.1 Theoretical expansion/contraction for reactions with calcium sulphate

Aluminate hydrate in hardened concrete	Product of reaction with calcium sulphate	Expansion
C_4AH_{13}	'Monosulphate'	1.1
C_4AH_{13}	Ettringite	2.6
C_4AH_{19}	Monosulphate	0.8
C_4AH_{19}	Ettringite	1.9
C_3AH_6	Monosulphate	2.1
C_4AH_6	Ettringite	4.8
$Ca(OH)_2$	$CaSO_4$	2.2

From consideration of the data in Table 2.1 there is some doubt that the expansion arising from the reaction of the calcium aluminate hydrates with sulphate is the sole cause of the damage readily shown in laboratory experiments. Nevertheless cements with low C_3A contents, the sulphate-resisting Portland cements, are notably more resistant to sulphate attack.

Sulphate-resisting properties are exhibited by concretes incorporating pfa or ggbfs as part of their cementitious content. Such concretes have a lower free calcium hydroxide content and the C_3A from the Portland cement is effectively diluted. It is also likely that the chemistry and physics of the paste are modified to inhibit the ingress of the sulphate and to change the nature of any reaction products.

When excessive amounts of sulphate are incorporated in the fresh concrete the reaction with C_3A may be continued after the cement has set and begun to harden. If the expansion caused is greater than the tensile strength of the paste, disruption occurs; the disruption will be extensive because of the wide dispersal of the sulphate. However, if the paste can cope with the forces, expansive or shrinkage-compensated properties can be achieved.

When external sources of sulphate are in contact with hardened concrete they can react with the outer surface, but if the concrete is porous they may be absorbed to react further into the body of the concrete. In practice, therefore, sulphate attack from external sources is manifested as a progressive weakening and disintegration of the surface. With very porous materials it is possible for the sulphate to be absorbed to such an extent that disruption takes place more generally but such a mechanism is normally only observed in laboratory experiments. As a consequence, concrete which has been damaged by contact with sulphates can be repaired by removal of the outer weakened layers until sound concrete low in sulphates has been reached.

Frost attack

As water turns to ice at its freezing point of 0 °C, there is an increase in volume of about 9%. When porous material is saturated with water this expansion on freezing can lead to disruption of the material if its strength is insufficient to resist the forces involved. Unless the material is more than 92% saturated, however, disruption does not occur since there is empty space to accommodate the expansion. The saturation need not be over the whole volume of the material and surface layers can suffer frost attack even though the main body is unaffected.

In saturated concrete the water filling the gel pores will not freeze under normal conditions because its freezing point is far too low. This is because the liquid in the very fine gel pores is under high surface tension forces which prevent the formation of the nuclei of ice responsible for initiation of ice growth. Water in the capillary pores freezes and as the ice grows it forces unfrozen water into the surrounding gel pores. Since a particular section of gel will be surrounded by capillaries all trying to push water into it, the hydraulic pressure build-up can be sufficient to disrupt the gel system.

Although the pore water in concrete contains dissolved salts the ice formed on freezing is pure H_2O. The remaining unfrozen liquid in the capillaries therefore becomes more concentrated. It is being forced into gel pores already containing water which is at a lower concentration and an osmotic pressure develops as the two solutions in contact attempt to equalize. This pressure is in the opposite direction to the hydraulic pressure caused by the expanding ice and so the two act together to produce larger forces in the paste. Water-filled voids and saturated porous aggregate particles will behave in a similar manner. When porous particles or voids under aggregate particles are near a surface there will usually be pop-out.

Low water–cement ratios lead to a smaller capillary pore system so that the freezing effects are much reduced. Alternatively air entrainment, when effective, gives an extensive system of bubbles which are separated by about 0.25 mm of gel and are not water filled. When water in surrounding capillaries freezes, the hydraulic pressure build-up is accommodated by expansion into the air bubble system.

One episode of freezing is not enough to cause damage and major effects occur after a number of freeze-thaw cycles, particularly if they succeed each other rapidly. Since the exposed surface of concrete will be more susceptible to rapid cooling and thawing and because the surface is more likely to be saturated, the general effect of frost attack is a scaling or delamination of the surface.

A solution of a de-icing salt in water has a lower freezing point than pure water. Also, a mixture of ice at 0 °C with a solution of de-icing salt leads to a rapid temperature fall as the ice melts. The melting ice dilutes the salt solution and the diluted solution has a different, higher, freezing point. This continuous process of change in solution concentration occurs in small volumes of paste and the effect is to set up a great many rapid micro freeze-thaw cycles with a consequent increase in damage. All de-icing compounds will cause the same effects to a greater or lesser degree depending on their efficiency as de-icers.

Salt crystallization

The word 'salt' is used in chemistry to denote the products of neutralizing an alkali with an acid and should not be confused with the common name 'salt' for sodium chloride. Thus magnesium sulphate, potassium nitrate and calcium chloride are also 'salts'. They dissolve in water to an extent depending on the actual solubility of the salt and the temperature. Thus a hot saturated solution of sodium chloride will deposit crystals of sodium chloride on cooling.

If solutions of salts move through concrete, water can evaporate when the solution reaches an exposed surface. If the evaporation rate is fairly slow it will tend to take place on the surface, leading to surface deposits; this occurs in mild climates as lime bloom or efflorescence. If the evaporation rate is fast, as occurs in hot arid climates, evaporation can take place before the solution reaches the surface, and when the solution reaches a particular concentration crystals will deposit in the capillary pores of the paste. Some types of crystals will grow in contact with their saturated solution and this growth can exert pressures on the paste structure analogous to those induced by the freezing of water.

Magnesium sulphate is particularly aggressive in this respect and is used to measure the unsoundness of aggregates by a test involving saturation and crystallization (BSI, 1989). Although the mechanism appears to be similar to that of freezing, it cannot be identical since there is no good correlation between magnesium sulphate soundness and freeze–thaw susceptibility of aggregates.

Salt crystallization damage is restricted to surfaces where drying takes place but the effect can be so pronounced that large volumes of concrete can be destroyed.

Fire damage

The hydrates of cement lose water when heated and become dehydrated to an extent depending on the temperature. Crystalline materials lose their water at clearly defined temperatures so that, for example, calcium hydroxide loses water to give calcium oxide at a temperature of 650 °C. Non-crystalline materials such as C–S–H gel lose their water over much broader temperature ranges so that, for example, C–S–H kept at 400 °C for some hours may lose as much water as if it were heated to 600 °C for some minutes. The dehydration of the hydrates leads to a change in structure which can lead to significant loss of strength.

Aggregates can also change their structure during heating and in the case of limestone (calcium carbonate) there can be a complete disintegration at 850 °C when it changes to calcium oxide. With flints the change is accompanied by spalling. Also the thermal expansion of the aggregate will not necessarily be matched by shrinkage of the paste as it dehydrates so that the differential movements lead to a reduction in bonding between the two.

The low thermal conductivity of concrete results in a slow temperature build-up in the interior of concrete and much of the heat is used to evaporate free water. The damage is therefore restricted to the outer surface in contact with the heat unless this contact is prolonged. There is some controversy about the maximum continuous temperature which Portland cement concrete can withstand without significant damage but a consensus is probably in the range 200–300 °C.

Refractory concretes are made with HAC and an aggregate which will not change or decompose on heating and whose thermal expansion matches the contraction of the HAC paste as it dehydrates. When the temperature is such that dehydration of the paste is advanced, a different type of so-called ceramic bonding between alumina particles takes place.

2.2.4 External effects leading to corrosion of reinforcement

Unprotected steel exposed to water and oxygen will corrode. The corrosion product, rust, is one of a series of hydrated iron oxides, the composition depending on the conditions of corrosion. Its volume is several times greater than the volume of metallic iron from which it is derived.

$$Fe \quad \overset{O_2}{\underset{H_2O}{\rightarrow}} \quad Fe_2O_3.H_2O$$

Portland cement concrete gives very good protection to steel, two factors being involved. Cement paste is highly alkaline and, as corrosion starts, the product is immediately converted to an oxide which forms a dense film, a so-called passive layer, on the surface of the steel in contact with the paste. Additionally, a sufficient volume of low permeability concrete between the steel and the atmosphere reduces the ingress of oxygen to the steel.

When the latter condition is not met, either by an insufficient volume (i.e. low cover) or an appreciable permeability, or indeed both, the protection is rapidly lost. This is because normal air contains carbon dioxide which reacts with the hydroxides in the cement paste and converts them to carbonates which have a much reduced alkalinity and no longer form the passive film on steel. The passive film can also be destroyed by reaction with chlorides. These may be incorporated in the concrete at the time of manufacture or may enter the concrete from an external source at a later age.

Carbonation

Normal air contains 0.03% carbon dioxide gas (CO_2) and this reacts with hydroxides to form carbonates. The reaction is used as a method of measurement so that, for example, if a known volume of dried air is passed through an absorbent containing a hydroxide, the increase in mass of the absorbent is equal to the mass of CO_2 in the volume of air:

$$CO_2 + 2NaOH \rightarrow Na_2CO_3 + H_2O$$
$$CO_2 + Ca(OH)_2 \rightarrow CaCO_3 + H_2O$$

Sodium carbonate is soluble and is alkaline, although to a lesser extent than sodium hydroxide, whereas calcium carbonate is insoluble. There is normally insufficient sodium in the paste for the sodium carbonate to give enough alkalinity to passivate steel.

Air is in direct contact with the external surfaces of the concrete and the surfaces will be carbonated very quickly after first exposure. As air enters the capillary pores of the concrete the CO_2 it contains is rapidly used up by reaction with the hydroxides. There is therefore decarbonated air in the concrete in intimate contact with normal air outside. In this condition there is a positive movement of CO_2 towards the depleted air in an attempt to restore a uniform concentration. This movement from a high concentration to a low concentration is called 'diffusion'.

Because CO_2 reacts rapidly and quantitatively with the hydroxides it cannot move past the barrier of uncarbonated concrete until all the hydroxide is converted, at which time it begins to move into the next increment of uncarbonated concrete. For this reason the carbonated layer is fairly sharply defined.

As carbonation proceeds the free calcium hydroxide is depleted. C–S–H gel is only stable in the presence of free $Ca(OH)_2$ so in the absence of free $Ca(OH)_2$ it decomposes to release calcium hydroxide leaving behind a hydrated silica skeleton. The other hydrates behave similarly and so the carbonated layer is a framework of silica, alumina and iron oxide filled with calcium carbonate. This is strong and less permeable than the original concrete.

As the carbonated layer becomes thicker, the distance that the CO_2 must diffuse through to meet hydroxide is increased and the carbonated zone itself gives more resistance to gas flow. The increase in carbonation depth is therefore exponential. One model proposed (Kondo *et al.*, 1986) for the variation of carbonation depth d with time t is

$$d_{(mm)} = At^{1/2}$$

A is a measure of concrete quality and is greater for poor concrete. Herein lies the problem of carbonation: it improves the quality of the paste by reducing permeability but a significant depth of carbonation in a relatively short time indicates poor quality concrete and any improvement caused by reaction with CO_2 is insufficient to bring the concrete up to good quality.

Carbonation is affected by the moisture condition of the concrete in two ways. Dry calcium hydroxide reacts very slowly with CO_2 and the reaction is much faster and more complete when a surface film of water, saturated with the solid, is present on the grains of the hydroxide. Very dry concrete therefore carbonates at a slower rate than damp concrete. On the other hand, if the pores of the paste are completely filled with water the CO_2 gas must diffuse through the water and the diffusion rate is considerably less than that through air.

When the depth of carbonation has reached embedded metal the paste in contact with the metal will have a much reduced alkalinity and the passive or protecting oxide layer will no longer form in the presence of oxygen.

Chlorides

The passive oxide film formed on steel surfaces in the presence of hydroxide ions can be destroyed by high levels of ions which form soluble iron compounds. Chlorides are abundant and are the most common cause of such depassivation.

The C_3A phase in Portland cement reacts with the sulphate in the cement to form sulphoaluminates and has been discussed in section 2.2.3. Chloroaluminates can also be formed and the chloride in these is rather insoluble in pore liquid. Cements containing high levels of C_3A can therefore 'bind' some chloride into an insoluble and therefore non-reactive form. Some of the chloride incorporated in fresh concrete as an admixture or from an unwashed marine aggregate can therefore be immobilized to some extent if there is sufficient C_3A hydrate present. It appears, however, that if sulphate and chloride are competing for reaction with C_3A, the sulphate is more likely to win.

Carbonation of chloroaluminates releases the bound chloride into a free mobile form. Chlorides in the outer carbonated zone of concrete can therefore be washed away or alternatively washed into the uncarbonated concrete.

Chloride entering concrete from an external source is to some extent analogous to carbonation. If water with a high concentration of chloride is in contact with concrete saturated with chloride-free water diffusion of chloride ions into the concrete will take place in an attempt to produce a uniform chloride concentration in the water. With concrete saturated with seawater, for example, the chloride concentration of the internal water will be lower, because of reaction with the C_3A, than that of the body of water external to the concrete and chloride diffusion into the concrete will take place. Of course if concrete saturated with seawater is in contact with fresh water then free chloride ions will diffuse out of the concrete.

The diffusion of chloride will be through the water in the capillary pores and, the finer and less continuous the pores, the slower will be the diffusion rate. In practice the ingress of chloride into concrete by diffusion is a relatively slow process and cycles of drying followed by saturation is a much more potent mechanism for the ingress of chloride.

Overall, dense concrete of high alkalinity and low permeability can carry higher levels of chlorides without the occurrence of corrosion than permeable carbonated concrete. The risk of reinforcement corrosion associated with carrying levels of chloride content in both uncarbonated and carbonated concrete is summarized in Table 2.2 (Pullar-Strecker, 1987).

Table 2.2 Corrosion risk in concrete containing chlorides (Pullar-Strecker, 1987)

Chloride (per cent by weight of cement)	Condition of concrete adjacent to reinforcement	Corrosion risk
Less than 0.4%	1. Carbonated	High
	2. Uncarbonated, made with cement containing less than 8% tricalcium aluminate (C_3A)	Moderate
	3. Uncarbonated, made with cement containing 8% or more C_3A in total cementitious material	Low
0.4%–1%	1. As above	High
	2. As above	High
	3. As above	Moderate
More than 1.0%	All cases	High

Where sulphate is present at more than 4.0% by weight of cement, or where chloride has entered the concrete after it has hardened (rather than being incorporated at the time of mixing), the risk of corrosion is increased in all cases.

References

BRE (Building Research Establishment) (1968) Shrinkage of natural aggregates in concrete, *Digest 35* BRE, Garston.

BSI (British Standards Institution) (1989) BS 812: *Testing Aggregates*. Part 121: *Method for Determination of Soundness*, BSI, London.

Concrete Society (1987) Concrete core testing for strength, *Technical Report 11*, Concrete Society.

Harrison, W.H. (1987) Durability of concrete in acidic soils and waters, *Concrete*, **21** (2) 18–24.

Hawes, F. (1986) *Appearance matters — 6: The Weathering of Concrete Buildings*, Cement and Concrete Association, Slough.

Kondo, R., Daimon, M. and Akiba, T. (1968) Mechanism and kinetics of carbonation in hardened concrete, *5th International Symposium on the Chemistry of Cement*, vol. 3, pp. 402–8, Tokyo.

Lea, F.M. (1970) *The Chemistry of Cement and Concrete*, 3rd edn, Edward Arnold, London.

Neville, A.M. (1977) *Properties of Concrete*, 2nd edn, Pitman, London.

Pullar-Strecker, P. (1987) *Corrosion Damaged Concrete, Assessment and Repair*, Butterworths, for Construction Industry Research and Information Association, London.

Structural investigations   3

A.F. Baker

(Formerly STATS)

3.1 Introduction

Before undertaking repairs to concrete structures it is essential to discover the cause(s) of distress or deterioration. The importance of a correct and comprehensive diagnosis cannot be overstated and, when considered in relation to the cost of many repair contracts, the cost of initial investigations and the preparation of concise specifications is minimal.

Reinforced and prestressed concrete should not be regarded as a 'maintenance-free' material. The durability and maintenance requirements of concrete structures should be related to an anticipated design life which incorporates a planned maintenance programme. Concrete structures can be designed to give a useful life in excess of 100 years and the material itself can remain sound after hundreds of years' exposure. However, concrete can also fail dramatically after only a few years. Unfortunately, unlike many other factory-produced construction materials such as steel and brick, identical ingredients of cement, water and aggregate with only small changes in mix design, water–cement ratio etc. can produce either an exceptionally durable material or a low quality porous permeable rubble.

Well-designed, compacted and cured reinforced concrete can be a very durable material. However, many of our buildings and structures were constructed at a time when concrete was regarded as a low maintenance or maintenance-free material which would be tolerant of shortcomings in mix design, placing and curing. In addition it was not realized how susceptible poor quality concretes were to deterioration due to poor aggregate selection, attack by de-icing salt or chloride-containing admixtures, atmospheric carbonation etc. There is therefore a continuing and increasing need to undertake the inspection of many of our concrete structures where premature deterioration has occurred, both to determine the need for any repair work and to determine whether additional protection and/or maintenance may be necessary. The purpose of inspection therefore is not only to document the type(s) and extent of deterioration identified but also to investigate the cause(s) of the deterioration and provide guidance on appropriate remedial action to prevent or mitigate against its reoccurrence (Fig. 3.1).

Fig. 3.1 Deterioration of repair to calcium-chloride-affected concrete after four years. Incorrect diagnosis and specification.

Problems with the lack of durability of many concrete structures arise because the engineer has insufficient knowledge of the life of his materials and the effectiveness of good detailing, as well as the influence on construction of the quality of workmanship on site. A building or bridge showing signs of spalling concrete after 20 years may not be due solely to poor workmanship or materials but may also be a result of a lack of knowledge of material durability under actual exposure conditions. All construction materials deteriorate in time and problems stem mainly from the premature unforeseen deterioration of materials and structural components.

In planning an investigation, therefore, it is important to define the scope and purpose of the investigation as these aspects are not always self-evident. It is also important to bear in mind that inspection and testing work can be very time consuming and due allowance should be made for this.

3.2 Planning an investigation

3.2.1 Preliminary considerations

A site visit to inspect the structure is recommended before planning an investigation. However, in certain circumstances, where a bridge inspection follows on from a principal inspection for example, this may not be necessary. The initial visual inspection is normally undertaken by the client's representative and the consulting engineer/materials scientist nominated to conduct the investigation. It is at this stage that both the client and the consultant should clarify and agree on the scope and objectives of the inspection. Factors which need to be considered include the following:

1. Is the integrity of the structure or any element of the structure in doubt? Is this part of the initial brief or may there be a requirement for examining integrity if repairs are needed?
2. Do adequate drawings exist of the as-built construction including rebar sizes, spacings, detailing etc. or is there a requirement to determine steel layout/concrete strength now, or prior to any repair being required?
3. Is access to parts of the structure likely to be difficult, involve expensive equipment or be restricted in terms of time? Is there any deadline by which the work must be completed (i.e. budget requirements, other maintenance requirements, change of ownership etc.)?

   It is normally more cost effective to undertake inspection and testing work in stages with a review of findings at the end of each stage. However, this is often restricted by the availability and expense of access, tight timescales, budget constraints etc.
4. What are the requirements for personnel to undertake the work, i.e. materials engineer/structural engineer/geotechnical engineer etc.?
5. Is the structure currently in use and what restrictions are there on time of work, use of noisy equipment etc? Are power and water available?
6. Is it necessary to define the extent of deterioration for budget-estimating purposes or in the preparation of approximate Bills of Quantities?
7. Are recommendations on repair options or preventative maintenance/ additional protection required, and is any order of priority to be given to this?
8. What further life is the client anticipating for the structure and is there a requirement for the investigation of any possible latent defects in otherwise sound areas?

Documents and drawings should be examined before carrying out inspection work on site if this is possible. Information of value would include the following:

1. age and details of any previous remedial/maintenance work;
2. construction details including foundations, joints, reinforcement etc;
3. details of concrete, especially specifications, mix designs and materials sources;
4. records of any previous inspections such as principal inspections of bridges;
5. scale drawings, general arrangements, plans and elevations.

The environment to which the structure is exposed should also be assessed and if possible the structure should be examined during dry and wet weather to ascertain whether joints and drainage are working properly (Fig. 3.2). When planning an investigation it should be remembered that certain areas of structures may be more susceptible to damage due to frost action or alkali–silica reaction (e.g. foundations). An examination of the layout and effectiveness of any waterproofing or joint-sealing systems used should also be included in the inspection programme.

Fig. 3.2 Failure of drainage system on bridge pier/abutment.

On large inspection contracts or where access is difficult or restricted in time it is always advisable to construct and agree a timed programme of work with the client and, where applicable, a specification for the work should be compiled to include, for example:

1. survey objectives;
2. plans of the structure;
3. an agreed system of proforma sheets, reporting format etc.;
4. site and laboratory work programmes.

3.3 Test and inspection techniques

A considerable range of test and inspection techniques is available for use *in situ* during inspection, or on samples removed from the structure for laboratory investigations. Test and inspection techniques can be subdivided into the following:

1. those dealing with structural integrity or location of reinforcement/voidage, e.g. acoustic emission, radiography, covermeter surveys;
2. those dealing with concrete quality and composition, e.g. core testing for strength, Schmidt hammer testing, determination of depth of carbonation;
3. those dealing with prestressing steel or reinforcing steel serviceability and condition.

In turn, test and inspection techniques can be subdivided into *in situ* and laboratory methods. The majority of the test techniques are described in appropriate British or American Standards or in technical publications, conference proceedings and papers. A number of references are included in this chapter which should be consulted for more detailed information on various test and inspection methods. A summary table of methods is also included (Table 3.1).

3.3.1 *In situ* sampling and testing

A very wide range of *in situ* sampling and test techniques is available for the inspection of concrete structures. Some of the test methods have been developed specifically for concrete inspection purposes, and others have been adapted for use on concrete although the techniques have been principally developed for the inspection of other materials, e.g. steel. Many of the more sophisticated techniques require specialist equipment and expertise, particularly in assessing results, and are only infrequently used during routine condition surveys. However, the inspection and testing of concrete structures, particularly by non-destructive methods, is a rapidly expanding and advancing field and the use of sophisticated equipment is likely to become much more widespread.

As stated earlier, inspection techniques can be subdivided into three categories:

1. determination of structural integrity;
2. determination of concrete quality;
3. determination of steel serviceability and condition.

The test and inspection methods under these categories are given in Table 3.1 which gives references to where more detailed information may be found. The more commonly used test and inspection methods are as follows:

Table 3.1 *In situ* test and inspection techniques.

Test method	Principal reference	Application and layout	Property assessed	Remarks
Structural integrity and layout				
Visual survey including (a) crack mapping (b) endoscope survey (c) photography, video (d) stereo pairs	Bridge Inspection Panel (1984) ACI (1968) Manning and Bye (1983) Cement and Concrete Association (1988)	All elements and structures	Condition and monitoring	Simplest but most important inspection method
		All elements and structures	Condition and monitoring	Include scales and colour charts as appropriate Close-ups and general views
Hammer testing Chain drags etc.	Moore *et al.* (1973) Savage (1985) Manning and Bye (1983) Cantor (1984)	All elements and structures	Presence of cracks, spalls and delaminations	Hammer test or chain drag may only detect surface laminations to 75–100 mm deep
Covermeter surveys to locate reinforcement, prestressing ties etc.	BSI (1988b) BSI (1986a)	All elements and structures	Location and concrete cover Bar sizes also determined	Influenced by magnetic aggregates and difficulties with lapped or closely spaced bars or layers of bars; modern instruments much more effective at locating steel up to 300 mm deep
Thermography	Manning (1985) BSI (1986a) Manning and Holt (1980) Kunz and Eales (1985)	Principally bridge decks (asphalt and plain concrete) and other elements	Presence of laminations	Influenced by water on deck and prevailing weather; rapid scan system
Radar	Manning (1985) Cantor (1984) BSI (1986a) Kunz and Eales (1985)	Principally bridge decks	Voidage etc. on large scale including reinforcement, ducts etc.	Bulky equipment; signals can be difficult to interpret
Acoustic emission	Mkalar *et al.* (1984) Hendry and Royles (1985) Manning (1985) BSI (1986a)	All structures during load testing	Determination of initiation and origin of cracks	Specialist inspection equipment and interpretation
Dynamic response e.g. sonic echo, continuous vibration	Stain (1982) Manning (1985) BSI (1986a)	Principally pile testing	Pile integrity	Specialist inspection equipment and interpretation
Load testing of structures	BSI (1986c) Menzies (1978) Jones and Oliver (1978) ACI (1985a)	Structures/elements	Deflection under loads against structural analysis	Expensive but informative

Determination of concrete quality and composition

Method	References	Applicability	Measures	Comments
Core and lump samples	Concrete Society (1987), BSI (1981), ASTM (1987a)	All elements and structures	Used in physical, chemical and petrographic analysis of quality	Required for almost all laboratory testing; restricted by access to certain members; can be expensive
Power drilled samples	Building Research Establishment (1977)	All elements and structures	Chloride, sulphate and moisture content of concrete	Simple but subject to errors of depth/cross contamination and sampling
Partially non-destructive assessment of strength i.e. Windsor probe, internal fracture, pull-out testing, Schmidt Hammer etc.	BSI (in preparation), BSI (1981), Keiller 1982	All elements and structures	Strength, usually converted to equivalent compressive	Varying in simplicity; often wide margin of error (15%–30%) depending on method and availability of calibration. Information on cover concrete only. Often operator and equipment sensitive. Can damage concrete
Ultrasonic pulse velocity testing	BSI (1986c), BSI (in preparation), ASTM (1987c)	All elements and structures	Concrete quality uniformity, presence of cracks, voids, weak layers etc.	Rapid scan technique through thickness of concrete; can be correlated with strength
In situ permeability tests, e.g. ISAT, Figg test, Clam	Concrete Society (1988), Lawrence (1981), Montgomery and Adams (1985)	All elements and structures	Concrete quality and permeability, generally restricted to cover	Can be difficult to use in situ and slow; may only give a permeability index measurement, not true intrinsic permeability, and can be influenced by moisture content of concrete, surface condition etc.

Steel serviceability and condition

Method	References	Applicability	Measures	Comments
Half-cell potential mapping	ASTM (1987f), Figg and Marsden (1985), ASTM (1983), Baker (1986)	All elements and structures	Likelihood of corrosion of reinforcement (and possible rate)	Single-cell and two-cell methods used with copper, silver reference electrodes. Requires in situ calibration
Resistivity of cover	Manning (1985), Vassie (1980), Figg and Marsden (1985), Wenner (1915)	All elements and structures	Electrical resistance of cover concrete	Four-probe method used with either embedded or surface-contact electrodes. Measurement of circuit resistance may also be valuable

3.3.2 Determination of structural integrity

Visual surveys

Guidance on undertaking visual surveys of structures is not well-covered in the literature, particularly in the UK, although it is perhaps the most important and certainly one of the most useful and informative parts of any inspection programme. The American Concrete Institute have published a *Guide for Making a Condition Survey of Concrete in Service* (ACI, 1968) and the *ACI Manual of Concrete Inspection* which gives guidance on inspection techniques (ACI, 1981). Similar information is published by the Canadian Highways Authorities (Manning and Bye, 1983; Manning, 1985).

Visual surveys should be carried out by experienced personnel capable of making careful observations and recording findings in a systematic manner. Inspection should always be thorough and systematic, paying attention both to areas exhibiting distress and areas which are visually sound for comparison. A number of points should be considered before a comprehensive visual survey of a structure is undertaken or where the visual survey is to be used in conjunction with the results of *in situ* testing and sampling.

Time spent in preplanning a survey will frequently result in much more useful information being recorded on site. It is worthwhile making arrangements for the following in advance of site work.

Photography A good single lens reflex camera is essential, together with telephoto and macro lenses and films with speeds ranging from 100 to 1000 ASA. A powerful automatic flash gun will frequently be required for work within buildings or beneath structures. It is important that a proper scale should be included with photographs used for record purposes, as well as a colour chart, so that processing can be uniform. A camera with a databack (time and date recorded on film) is also helpful.

Photography is an excellent method of recording defects such as crack patterns etc. which can be marked on the structure with chalk. Taking photographic panoramas of structures and stereopairs is relatively simple and can be of great value in recording defects for later analysis.

Taking notes A small tape recorder is an invaluable tool for recording information on site where difficulties of access, weather etc. can make note taking very laborious and hence not very effective.

Proforma sheets/check lists During the preplanning stage it is often possible to prepare beforehand much of the paperwork necessary to record information on site. If plans and elevations of the structure are available a simple proforma sheet can be drawn up giving details of the grid referencing system, location sketch, visual survey key, photographic notes etc. A typical proforma sheet for a building inspection, together with a check list, is given in Fig.3.3.

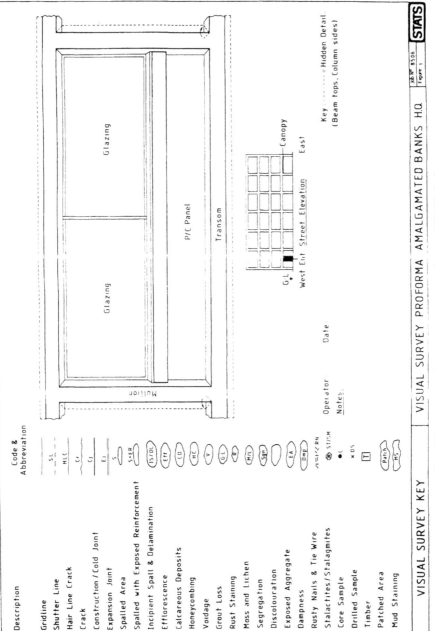

Fig. 3.3 Example of a visual survey proforma (STATS Ltd).

VISUAL SURVEY KEY | VISUAL SURVEY PROFORMA AMALGAMATED BANKS H.Q

Description	Code & Abbreviation
Gridline	S L
Shutter Line	MLC
Hair Line Crack	Cr
Crack	Cj
Construction / Cold Joint	Ej
Expansion Joint	S
Spalled Area	S+ER
Spalled with Exposed Reinforcement	IS/DL
Incipient Spall & Delamination	Eff
Efflorescence	CD
Calcareous Deposits	HKC
Honeycombing	V
Voidage	GL
Grout Loss	R
Rust Staining	M/L
Moss and Lichen	Sgn
Segregation	D
Discolouration	EA
Exposed Aggregate	Dmp
Dampness	RN
Rusty Nails & Tie Wire	ST/SM
Stalactites/Stalagmites	C
Core Sample	x DS
Drilled Sample	T
Timber	Patch
Patched Area	MS
Mud Staining	

STATS

Job No 8506
Figure 1

The visual survey is normally used as the basis for any on site testing, non-destructive testing and sampling. It is therefore very important to record as much detailed visual information as possible as this may be used later to explain, for example, the results of anomalous core test results or to provide reasons for high chloride levels or low cover readings on the structure. In addition it is important to record information with respect to some form of grid-referencing system on the structure so that visual information can be correlated with results of other tests and samples can be accurately located.

Tapping surveys

A small club hammer should be used to sound the surface of the concrete in order to detect incipient spalls and laminations over the reinforcement. On horizontal surfaces such as bridge decks a metal rod can be used to sound the surface or alternatively various types of chain drag can be used; one type consists of 0.5 m lengths of 25 mm chain attached to a T-shaped handle. The chain drag technique is believed to be less accurate than simple hammer soundings (Manning, 1985). It should be borne in mind that considerable force may be necessary to sound concrete for laminations where the cover to the steel is greater than about 40–50 mm. Although a club hammer will locate laminations accurately at this depth, beyond 100 mm sounding concrete can be ineffective in picking up deep laminations. More sensitive equipment has been used for locating deep laminations, including the 'Delamtech' (Manning, 1985) and the instrumented delamination device (IDD) (Savage, 1985). Infrared thermography has also been used (Manning and Holt, 1980).

Covermeter surveys

Covermeter surveys form part of most condition surveys on concrete structures. Surveys are generally one of three types:

1. to locate main and secondary reinforcement to determine bar sizes, bar spacings and cover to individual bars;
2. to determine minimum cover and cover variability across an element (generally the concrete surface is subdivided into 500 mm or 1 m grid squares and the minimum cover in each grid square is recorded);
3. to locate reinforcing steel, prestressing strand, cladding, ties etc. either for inspection or for avoidance during core sampling, direct pulse velocity surveys etc.

The use of covermeters is described in BS 1881 (BSI, 1988).

A wide range of new covermeter designs are now available offering significant improvements on older designs which were bulky and clumsy to use. Manufacturers' claims for these instruments are as follows:

Maximum depth of location of steel	100–300 mm
Bar size measurement	Dial-in or calibration graph (5–40 mm bars)
Influence of magnetic aggregate	Most include compensation circuitry to eliminate effects
Bundles of bars	Some models have high resolution probes
Audible tone and display	Usually incorporated and selectable
Non-ferrous metals	Usually located at reduced sensitivity
Weight (including case)	3–4 kg including probes

Although undertaking a covermeter survey is a relatively straightforward procedure the following points should be noted.

1. Modern covermeters are quite complex and some training in their use is necessary.
2. It is always advisable to break out reinforcement in a few areas to calibrate the covermeter and to check that the selected bar size is present. Magnetic particles in concrete aggregates or cement replacement materials may influence results if their presence is not known.
3. Bundles of bars, lapped bars and layers of bars all present problems.
4. Deep-seated bars beneath a lighter surface mesh can be very difficult to locate, for example where mesh reinforcement has been incorporated into a gunite repair or reinforced concrete or where mesh reinforcement has been incorporated into cladding panels and it is necessary to locate ties at the rear of the panel. Offcuts of tying wires in soffits can also cause problems.

Thermography, acoustic emission, dynamic response and radar

These test techniques all yield information on structural integrity and have all been successfully applied in specialist investigations on reinforced concrete structures. Further information on the test techniques can be obtained from the references at the end of this chapter and their advantages and disadvantages are discussed in BS 1881: Part 201 (BSI, 1986a).

Load testing of structures and elements

In most cases the load-carrying capacity of a structure or its individual members can be determined sufficiently accurately by using a combination of materials investigation and structural assessment methods. Occasionally, however, the results of such an exercise may be inconclusive, particularly if structural drawings and/or data are not available, in which case an alternative means of assessing loading capacity must be considered. A direct full-scale load test is often the only means of satisfying loading criteria.

On existing structures, load tests are often required where a change of use is anticipated. For instance, an area previously used as offices might be destined to accommodate warehousing. Similarly a load test can be used to prove the integrity of a structure following physical, chemical or fire damage. In new structures load tests are employed to check the load-carrying capacity of elements where poor workmanship or deficient materials have cast doubt on the quality of the construction.

Unless testing to destruction, it is usually not possible to determine the ultimate load capacity of a structure of an element. Load tests can only be used to check proposed maximum loadings and to ensure that elements of the structure are able to take the designed working load with safety.

It may be possible to load test individual structural elements under laboratory conditions if the units in question are sufficiently portable. However, such opportunities are rare and it is more likely that the tests have to be undertaken *in situ*.

Several factors will determine the choice of loading medium and include cost, convenience, access and accuracy. Amongst the loading mediums available are

1. certified steel weights
2. water
3. bagged sand
4. kentledge
5. hydraulic jacks

For small localized load tests, certified steel weights are probably the most suitable medium in that they are clean, easily portable and simple to apply. Their use in larger load tests, however, is less desirable because of their high purchase and/or hire cost and the large number of personnel required to handle them at the site of the test, particularly if access is poor.

For load tests of up to 1.0 tonne/m^2 over a large area water is a particularly suitable medium. Using standard plywood panels and a lightweight steel framework a bottomless tank can be constructed over the test area. After lining the tank with an impermeable membrane it can gradually be filled with water until the desired loading is achieved. The advantages of water are fivefold.

1. It is inexpensive.
2. It is accurate.
3. Uniform loading is ensured.
4. It is readily available and easily transported provided that there is

(a) a convenient supply in the vicinity;
(b) a drain to dispose of the water on completion of the test.
5. Tanks can be erected in relatively inaccessible locations.

Even if great care is taken the membrane used to line the tank occasionally develops one or two small leaks. If leakage is an important consideration then additional preventative measures should be taken or an alternative loading medium should be considered.

Bagged sand can be used as an alternative to water if appropriate. The sand can be delivered in bulk to the site of the load test and weighed into sandbags using portable scales prior to application. Although relatively inexpensive to purchase, its transportation to the site and from site on completion of the test and its movement from the point of delivery to that of application can be expensive.

Kentledge, either seel, iron or concrete, is used where large test loads are demanded, e.g. for *in situ* pile tests, and, as it invariably has to be transported and placed using a crane, its use is limited to external applications.

Hydraulic jacks are the most convenient means of applying test loads for the following reasons.

1. The equipment is compact and lightweight compared with the loads which can be applied.
2. Suitable jacks are relatively inexpensive to hire.
3. Loads can be applied gradually and accurately.
4. Retests can be performed simply and quickly.
5. The number of personnel required to undertake the test is small.

The main disadvantage associated with the use of hydraulic jacks is that a reaction load has to be provided, often in the form of kentledge, which may not always be convenient and is almost certainly expensive. Jacks also provide a concentrated load and, in many cases, a distributed load is required. In some circumstances a means of reaction against other temporary or permanent structural members can be found.

Having chosen the loading medium, the manner in which it is applied and the method of monitoring its effect must be considered.

British Standards give guidance on the load testing of concrete structures (BSI, 1985) and their individual components. It is possible that in some cases the recommendations in the British Standards may not be entirely appropriate. Divergence from these recommendations may be necessary to suit the structure or member in question. In any event methods and procedures must be agreed by all the parties involved before carrying out the test.

Because the strength of the structure is suspect it is possible that the performance and behaviour of the structure will be unpredictable during the load test. For this reason safety supports must be installed beneath the structure if the load is to be applied vertically. These supports should be capable of taking the full test load together with the dead weight of the structure. So as not to interfere with the behaviour of the structure during the normal course of the test a working clearance should be provided between the structure and the safety supports. Scaffold should also provide for the dynamic effect of a falling load in case of sudden failure. A load falling through 50 mm may be equivalent to a static load of twice the nominal magnitude.

Advice on the maximum value of the test load can often be found in the appropriate British Standards Code of Practice and is usually expressed as an additional proportion of the maximum working load. The value of the test load can vary from these recommendations if it is felt appropriate.

Usually only a part of the structure will be loaded and it is necessary to make allowance for load sharing to ensure that the tested area is carrying the intended test load.

The behaviour of the structure can be monitored by measuring its deflection under load at points of importance. The number of locations will vary according to the nature of the structure but should include those of expected maximum deflection, i.e. at centre spans of beams and slabs or at the free end of cantilevers, and minimum deflection, i.e. adjacent to supports, fixings etc.

Deflections can be measured in many ways, of which the most popular are as follows.

Deflection dial gauges These provide a direct measure of movement up to a maximum displacement of 50 mm to an accuracy of 0.01 mm. Although they are simple to install and monitor they provide only single readings at given intervals during the course of the load test. Dial gauges are normally used for short-term load tests when continuous attendance is feasible.

Linear voltage displacement transducers These are capable of continuously monitoring deflection over the same range and to the same degree of accuracy as dial gauges and are normally used for dynamic and/or long-term load tests. The information they provide can be viewed at the time of the test on an oscilloscope or visual display unit, or stored on a data logger and recalled in the form of a time–deflection plot.

Electrical resistance and vibrating-wire strain gauges The operating range of these types of gauges is normally very small, although they can measure movement to a very high degree of accuracy. Consequently they are used not to measure the deflection of the structure but to measure the strain within the members themselves. Readings can be instantaneous or can be stored and reviewed over a period of time.

Plastic vernier strain gauges These provide instantaneous direct readings of movement over a large range and although they are inexpensive and simple to install their degree of accuracy is often poor. A cursor arrangement allows maximum and minimum readings to be recorded.

Demec gauges As with electrical resistance and vibrating-wire strain gauges, Demec gauges are highly accurate over a small range of movement. The change in distance between two metal reference points fixed to the surface of the material is measured using a device which comprises a dial gauge and a system of spring levers which magnify movement. Only instantaneous single readings are possible.

It is in most cases appropriate and advisable to apply the test load in a minimum of approximately six equal increments. After taking an initial reading of all gauges under zero test load the deflection of the structure should be monitored at each stage of the load test until the maximum load is achieved.

If creep of the structure under maximum load is an important considera- tion, then the test load should be maintained for an agreed period of time during which the deflection is monitored at regular intervals. At the end of this period, after reading all gauges, the load should be removed in similarly equal decrements, again whilst monitoring deflection.

In some structures, particularly those only recently completed, initial excessive deflection and creep during the maintenance of full load may be due to non-elastic settlement at the supports and in fixings. It is therefore advisable to carry out each load test at least twice so that a fair record of the structure's behaviour can be obtained.

The structure or member can be deemed to have passed the load test by checking one of the following:

1. comparison of the actual behaviour of the structure under load with the theoretically predicted performance (this is possible only if sufficient design data with respect to the materials used in construction are available);
2. comparison of recovery of the structure on completion of the load test with deflection under full test load (for this purpose acceptable percentage recoveries are given in the appropriate British Standards);

3.3.3 Determination of concrete quality

Core sampling

Coring is the most widely used method of sampling and is the only technique capable of producing samples suitable for use in most laboratory analytical and test methods. Cores taken from structures vary in diameter from 40 mm to over 200 mm. They are generally taken through the cover and for a depth into the body concrete at least equivalent to the diameter of the core.

The equipment available for core sampling varies considerably. Water- cooled diamond bits are normally used, driven by electric, air or hydraulic motors. Air and hydraulic motors are generally much more powerful than electric motors and are used particularly for inverted coring, into deck soffits for example, and where coring through steel cannot be avoided, for example drilling through bridge deck surfacing to sample the upper deck concrete for later chloride analysis. A water supply to the core bit is normally supplied by a small pump-pressurized canister. A steady flow of water is required in core drilling to prevent the barrel jamming due to overheating and to flush debris from around the cutting face of the core barrel. Typically, drilling one core 100 mm in diameter by 300 mm can require approximately 10 litres of water. In cold weather antifreeze can be used to prevent freezing of the water.

It is important to keep full and complete records of core samples removed from a structure for later analysis. Points which should be noted on site sample core record sheets include the following:

1. core diameter, length and number of pieces extracted (including steel);
2. which cracks in the core were caused during coring and which were present in the structure (an examination of the core hole can facilitate this; often when a core fractures the upper/outer portion revolves over the inner fixed portion obliterating any detail at the crack face);
3. an 'up' arrow marked on each piece of the core to indicate the direction the core was taken with respect to the outer surface of the concrete;
4. an orientation arrow on the outer face of the core to show the position with respect to any reference point, i.e. vertical datum or adjacent joint etc.;
5. a grid reference accurately locating the core on the structure;
6. a scaled photograph of the core location after core extraction is useful, to include any obvious features such as an area of cracking/dampness or an adjacent construction/expansion joint;
7. a sketch of the core to show how it can be reassembled in the laboratory, together with location of bars (these too should be numbered), cracks in the core hole etc.

An example of a typical core record sheet is given in Fig. 3.4. Cores are then normally marked with a sample reference number and wrapped in a material such as cling film. Cores should be transported to the laboratory in core boxes, or failing that in lengths of PVC drainpipe, to avoid further damage prior to inspection and analysis.

Appropriate British Standards (e.g. BSI, 1983b) and other guidance notes on testing for alkali–silica reaction (e.g. Cement and Concrete Association, 1988) include information on required core diameters and lengths etc. for specific laboratory test purpose, e.g. compressive strength, expansion testing for alkali–silica reaction etc. In cases where concrete strength is in dispute, guidance on sample locations in BS 1881: Part 120 (1983b), BS 6089 (1981) and Concrete Society Technical Report 11 (1987) should always be consulted.

SITE CORE RECORD

Client	Site	Job No	Site Ref
Date	Type of Rig	Operator	
Location of Core in Structure (Including grid refs)		Photograph (Tick) Core ☐ Hole ☐	

Direction of drilling (Tick box)	Method of flush (Tick box)	Dia of drill bit
Vertical up ☐ Vertical down ☐ Horizontal ☐ Inclined ☐ Approx angle ☐ +α° ☐ −α°	Water flush ☐ Air flush ☐ Dry ☐ Antifreeze ☐	

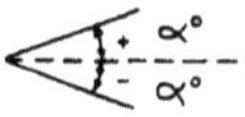

Depth of Carbonation		Number of Pieces	Reinforcement size
Max	Min	Assembled Length Max / Min	Distance to Reinforcement State Vertical or Horizontal

Operators Comments on CORE SAMPLE (state of Concrete, state of Steel, Cracks Delamination of Core etc)

Operators Comments on CORE HOLE (Cracks, Voids, poor compaction, Reinforcement, loss of water during coring etc)

Sketch of Core hole location (if necessary)

Notes:

1. Mark Core with site reference number.
2. All Core locations to be photographed or sketched.
3. All Cores to be labelled.
   (a) UP ↑ (b) OUT ← (c) 1 2 3
4. If outside of Core is destroyed by coring, mark Core with estimate of depth of new surface i.e

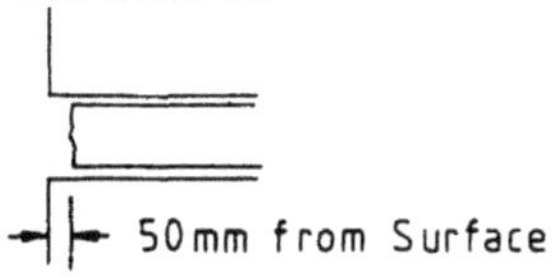

STATS

Fig. 3.4 Site core record sheet (STATS Ltd).

Powder-drilled samples

Powder-drilled samples of hardened concrete are normally taken in order to determine the chloride content of hardened concrete. Where chloride is suspected to have been derived from an external source, e.g. de-icing salt on a bridge, it is normal to take powder gradient samples at various depths, usually 5–25 mm, 25–50 mm, 50–75 mm, 75–100 mm etc., so that a chloride penetration profile with depth into the concrete can be plotted.

Gradient samples are also useful where it is suspected that calcium chloride has been incorporated into the concrete: the chloride content of concrete cast using a calcium-chloride-based accelerator remains constant and high with depth, whereas in a chloride-free concrete the 'base level' is generally practically zero unless a chloride-contaminated aggregate source or mix water has been used.

Powder sampling is described in Building Research Establishment *Information Paper IP 13/77* (BRE, 1977). Normally at least two holes are drilled with a 25 mm diameter masonry percussive bit and powder is collected at various depth increments for later laboratory analysis. The method is preferable to taking cores for chloride analysis as it is less expensive and time consuming, and samples can be taken down cracks etc. and in areas difficult to access with a core rig. Making good is also considerably easier.

Powder-drilled samples are also used for determining sulphate profiles in concrete. Their use in determining the alkali content of hardened concrete in suspected cases of alkali–silica reaction is not recommended, although powder samples may be useful in investigating local alkali concentrations, at expansion joints in bridges for example.

Powder samples have been used in the investigation of high alumina cement (HAC) to determine whether it is present, usually using the BRE chemical test (Bate, 1984). Powder samples are not recommended for degree of conversion determinations as the act of drilling the sample may frequently overheat it, giving misleading or erroneous results using differential thermal analysis. Small lump samples taken by drilling two or three small holes and breaking out the concrete between tend to give a clearer definition of HAC hydrate peaks and are preferred.

In situ *determination of depth of carbonation*

The normal method of determining the depth of carbonation of concrete is to spray a sample broken from the concrete surface, or the broken surface itself, with a solution of the indicator phenolphthalein made up as a 1% solution in alcohol–water. Phenolphthalein changes colour over the pH range 8.3–10 with the red colour on concrete visible above a pH of 9–9.5 (Parrott, 1987).

Although phenolphthalein indicator will give a measurement of depth of total carbonation, it does not provide information on the extent of partially or incompletely carbonated paste, which may extend to a greater depth into the concrete.

Carbonation testing should always be undertaken on a fresh fracture surface of concrete. Anomalous results can be obtained if the indicator is used on a powder-drilled sample because drilling tends to break up and expose the unhydrated cement grains normally present in concrete, giving a misleading reading (Pullar-Strecker, 1987). Drilling two or three holes and breaking out between to expose a fresh surface is normally the most suitable method.

Where the concrete is very porous, e.g. cast stone and certain lightweight aggregate concretes, a clear definition between carbonated and uncarbonated paste may be difficult to achieve. Often the indicator remains colourless, indicating that the concrete is completely carbonated, and then slowly develops a red colouration as lime is leached out of the concrete by the aqueous indicator solution. Where an accurate measurement of depth of carbonation is needed other methods, discussed in section 3.3.5 on laboratory testing and sample analysis, should be considered. The maximum depth of carbonation determined by phenolphthalein testing may underestimate the zone of reduced pH due to carbonation by as much as 20 mm (Parrott, 1987).

A description of the test method, together with a discussion on factors influencing the carbonation of concrete is given in BRE *Information Paper IP 6/81* (BRE, 1981).

Non-destructive assessment of in situ *strength*

A number of tests for assessing the strength of *in situ* concrete without taking and testing cores have been developed. All the tests are described in BSI (1986a) as 'near to surface tests' in that, in general, the strength of the concrete as a whole is estimated from tests performed on the cover concrete. Some of the more involved tests which require partial coring of the concrete, e.g. the Norwegian break-off tests, will estimate strength at deeper levels within the element being tested. A summary of the main test procedures together with estimates of precision and notes on application is given in Table 3.2.

A number of the test equipment kits available include information on calibration against the compressive strength of concrete, e.g. surface hardness, internal fracture and the Windsor probe. However, as can be seen from Table 3.2, the accuracy of strength prediction without a specific calibration is very poor, typically ±20%. According to Keiller (1985) it is unlikely that the accuracy of strength determination will be better than ±8 N/mm^2 at 95% confidence levels for normal strength concrete, even with a direct specific calibration, when using near to surface tests such as the Windsor probe, internal fracture and pull-off tests. In most cases where these tests are used no direct calibration

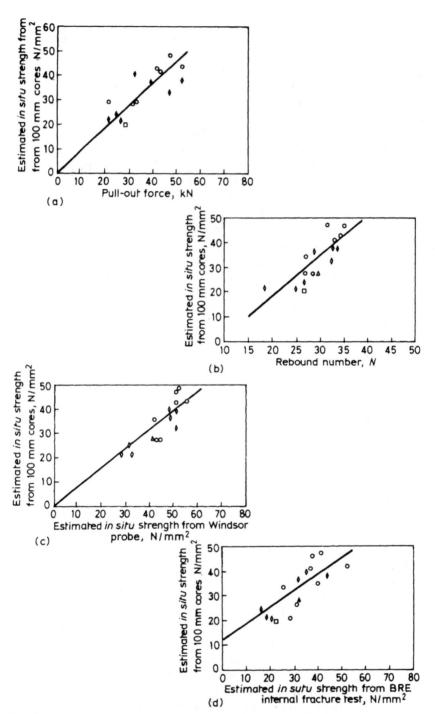

Fig. 3.5 Typical scatter of results for near to surface tests used in the determination of estimated *in situ* strength (after Keiller, 1982): relationship between estimated *in situ* strength from cores and (a) pull-out force in cast-in pull-out tests, (b) rebound number for first rebound, (c) from the Windsor probe test and (d) from the BRE test.

Table 3.2 Near to surface test methods

	Principal references	Compressive strength estimation of accuracy (from BS 1881: Part 201)	Depth of influence (mm)	Remarks
Surface hardness	BSI (1986b) Keiller (1982)	± 25% with specific calibration	30	Rapid test speed Carbonation affects results Specific calibration required for strength
Internal fracture	BSI (in preparation) Keiller (1982)	± 28% at 95% confidence with general calibration	17	Drilling required Specific calibration advisable Surface damage
Windsor probe test	BSI (in preparation) Keiller (1982)	± 20% at 95% confidence with specific calibration	25–75	Surface damage Calibration required
Pull-off test	BSI (in preparation)	± 15% accuracy on site with specific calibration	5 unless cored	Calibration required Partial coring advisable
Break-off test	BSI (in preparation) Keiller (1982)	Insufficient data	70	Surface damage
Pull-out test	BSI (in preparation) Keiller (1982)	± 20% if general calibration used ± 10% if specific calibration available	25	Drilling + under-reaming required Surface damage

is available. Near to surface non-destructive tests are therefore of little value in determining non-compliance of concrete with a specification, unless the concrete is clearly well below a specified strength (Fig. 3.5).

However, near to surface non-destructive tests do have several advantages.

1. Because non-destructive tests on concrete are simple and quick to perform, they can be used to determine concrete uniformity within a structure and can be an aid to quality control.
2. They are useful in the assessment of the level of strength in thin elements, e.g. HAC floor beams, or in areas where access for coring equipment is very difficult.
3. They are useful as an aid in deciding where best to take cores.
4. They are useful where surface properties of the concrete are important, i.e.
    (a) in the assessment of formwork stripping times;
    (b) in the assessment of fire damage or chemical deterioration of concrete;
    (c) in the assessment of surface durability, e.g. floor surface hardness testing for abrasion resistance etc.;
    (d) in the assessment of bonding of repair materials.

A discussion of the role of non-destructive testing of *in situ* concrete in structures is given by Tomsett (1981).

Ultrasonic pulse velocity testing

Ultrasonic pulse velocity testing of concrete is described in detail in BS 1881: Part 203 (BSI 1986c) and in ASTM C597 83 (ASTM, 1987c). The ultrasonic method of testing concrete depends upon the fact that the velocity of an ultrasonic pulse in concrete is related to both the density and the elastic properties of the concrete by the equation (Cement and Concrete Association, 1977)

$$V = \left[\frac{E_d}{\rho} \frac{(1-\upsilon)}{(1+\upsilon)(1-2\upsilon)} \right]^{1/2}$$

where E_d is the dynamic elastic modulus, ρ is the density, υ is the dynamic Poisson ratio and V is the velocity. As E_d is closely related to concrete strength, V can be used to estimate strength. The measured velocity is also influenced by defects in the concrete as shown in Fig. 3.6.

Methods of conducting pulse velocity surveys are given in BS 1881: Part 203 (BSI, 1986c). Three basic on-site test methods are described:

1. direct transmission with transducers on opposite faces of an element — the pulse passes through both cover concrete and body concrete;

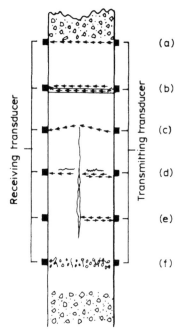

Fig. 3.6 Factors affecting pulse transmission through concrete (after Manning, 1985). (a) Sound concrete: sound waves travel the shortest distance between transducers. (b) Reinforcing steel: sound waves travelling through steel arrive before those through concrete. (c) Shallow crack: sound waves travel around the crack and transit time is increased; apparent pulse velocity is reduced. (d) Narrow crack: sound waves are partially reflected and partially transmitted with a large loss in amplitude but only a slight increase in transit time. (e) Wide crack: sound waves are wholly reflected; no signal received. (f) Voids and microcracks: sound waves may travel around or across (if water filled) with increase in transit time; apparent pulse velocity is reduced.

2. semi-direct transmission with transducers on surfaces at 90° to each other — the pulse passes through both cover concrete and body concrete;
3. indirect or surface transmission with both transducers on one surface — the pulse passes through the cover concrete only.

Method 1 is to be preferred wherever possible and is suitable for the determination of concrete uniformity and for the detection of large voids, discontinuities and cavities in concrete. It is suitable for the estimation of *in situ* strength if a suitable calibration chart is available. Where only a limited number of cores is available for correlation the formula

$$F_c = kV^4$$

where F_c is the compressive strength of the core (N/mm^2), k is a constant for that particular concrete and V is the pulse velocity (km/s), can be used to estimate strength from pulse velocity readings where the range of strengths to be expected is small (Cement and Concrete Association, 1977). The accuracy of the estimated *in situ* strength of concrete at a single location can be of the order of ±20% if a correlation is available for that particular concrete. Without a direct correlation the determination of strength from pulse velocity results is very poor.

Method 3 is the least sensitive transducer arrangement and for a given path length produces, at the receiving transducer, a signal which has an amplitude of only about 2%–3% of that produced by direct transmission (BSI, 1986c). The use of an exponential receiving transducer which has a built-in preamplifier may be necessary to boost the receiving signal.

The indirect test method can be used to determine the depth and orientation of air-filled cracks and is also a means of determining the quality of cover concrete with respect to the body concrete behind reinforcement, e.g. in the assessment of fire damage. Using the indirect method it is normal to construct a graph of transducer spacing versus pulse transit time. Discontinuities in the line indicate cracking normal to the pulse path or lower quality surface layers.

The sensitivity of method 2 is between that of methods 1 and 3.

When using these three methods care must be taken to avoid any reinforcement, particularly parallel to the path length, and it is suggested that a covermeter survey should be conducted prior to pulse velocity testing so that reinforcement can be avoided.

Pulse velocity measurements have been used to monitor strength growth in predicting striking times for shuttering, in determining release times for prestressing wires and in determining the depth and extent of fire and chemical attack on concrete members. Pulse velocity measurements are also used in the investigation of depth and orientation of air-filled cracks in concrete.

In the assessment of uniformity of concrete construction it has been suggested that, for normal standards of construction, the coefficient of variation of measured pulse velocity should be 1.5% or less for one load of concrete, 2.5% for several loads and up to 6% for an entire building constructed to good standards of workmanship (Tomsett, 1981). Pulse velocity measurements have also been used in the assessment of cracking and development of microcracking in structures affected by alkali–silica reaction (Cement and Concrete Association, 1988).

In situ *permeability tests*

A number of surface test methods have been used to assess the permeability

Table 3.3 In situ permeability test methods

Test method	Advantages	Disadvantages
ISAT	Non-destructive Considerable amount of comparable data available; covered by British Standards	Surface absorption method only giving values in millilitres per square metre per second; strictly a comparative method Sealing very difficult and moisture content/surface texture can be a problem
Figg	Air or water can be used Considerable amount of comparative data available	Values given in seconds for a pressure drop/increase; moisture influences results
Clam	Non-destructive and simple to seal Intrinsic permeability quoted	Adhesive needs time to cure Preplanning

Table 3.4 Typical permeability values for concrete

		Concrete permeability value		
		Low	Average	High
ISAT (ml/m²s)	10 min	<0.25	0.25–0.50	>0.50
	30 min	<0.17	0.17–0.35	>0.35
	1 h	<0.10	0.10–0.20	>0.20
	2 h	<0.07	0.07–0.15	>0.15
Figg absorption, water 50 mm dry concrete (s)		>200	100–200	<100
Modified Figg air, 55–50 kPA (s)		>300	100–300	<100
Clam intrinsic (m/s)		$<10^{-12}$	10^{-12}–10^{-10}	$>10^{-10}$

of *in situ* concrete. The most familiar test method is the initial surface absorption test (ISAT) method described in BS 1881: Part 5 (1970) currently being reviewed as BS 1881: Part 208. Other test methods include the Figg test (Concrete Society, 1988) and the Clam test (Montgomery and Adams, 1985). A comparison of the three test methods with their advantages and disadvantages is given in Table 3.3. Other *in situ* test methods are discussed in the Concrete Society report.

All of the methods discussed are generally used on a comparative basis only, although the ISAT method has been used in compliance testing on precast cladding panels and also in specifications for concrete repair materials applied on site. The Concrete Society has published typical permeability values for three grades of concrete quality for the ISAT, the Figg test and, by inference, the Clam test. Values are given in Table 3.4. However, the figures in Table 3.4 should be used for guidance only and not for any specification or contractual arguments on quality. Differences in the method used to condition concrete or in the test technique can produce significant differences in typical test values.

3.3.4 Steel serviceability and condition

Corrosion of reinforcing and prestressing steels in concrete is perhaps the most frequently observed symptom of distress seen on concrete structures. When reinforcing bars corrode it is not normally the loss in mechanical strength and steel cross-sectional area which is the initial problem, but the spalling and cracking of the concrete which results from corrosion of the reinforcement. This may both weaken the structure and present a public hazard on buildings or bridges, for example. Corrosion of prestressing strand, bar and wire can be much more structurally serious, owing directly to loss in section of highly stressed steel.

Various test and inspection techniques are used to investigate the corrosion of steel in concrete, although a number are laboratory-based techniques not yet in general use in the field (Berke, 1986). The most widely used *in situ* test methods are half-cell potential mapping, and the resistivity testing of concrete.

Half-cell potential mapping

The half-cell potential mapping of concrete can be used to delineate anodic (corroding) and cathodic (passive) areas of reinforcement. The technique was pioneered in the USA and is covered by ASTM C876-87 (1987f).

It is normally not possible to measure the potential of steel in concrete directly as it is remote from the external exposed surface of the element or structure. The potential developed by corrosion current flow, measured at the surface of the concrete, is therefore the expressed potential and will be modified by the moisture content, salt content and thickness of the cover. The potential recorded is normally a mixed potential resulting from both anodic and cathodic corrosion processes.

The basic test equipment is described in ASTM C876-87 (1987f) and has traditionally consisted of a copper/copper sulphate half-cell connected through a high impedance voltmeter into the steel reinforcement. Copper sulphate cells have a propensity for contamination and silver/silver chloride cells are now commercially available for survey work. The half-cell is then contacted to the surface of the concrete via a sponge or porous plug and contact fluid (normally water with detergent or alcohol).

The electrical potential of steel in concrete is a relative value and it is therefore possible to map out potential differences using two methods (Figg and Marsden, 1985). Method A involves a direct connection into electrically continuous steel using a single cell. Absolute potentials are recorded. Method B involves using two reference electrodes, one as a static probe and the other as the mapping probe. In this system absolute potential values are not recorded. In general method A has been preferred although, as the pattern and gradient of potential is more informative than the absolute value, method B should give comparable information.

In use on site the half-cell probe is used on a grid system with potentials being recorded at grid nodes (Fig. 3.7). The grid intervals chosen depend on the variability of readings at adjacent points. A grid spacing of 500 mm is normally sufficient, reducing to 250 mm in areas of intense corrosion activity. Grid spacings of 100 mm may be used where, for example, corrosion is localized at cracks.

The test method is simple to use and lends itself to automation via data loggers and the computer plotting of results (Baker, 1986). Half-cell potential data are normally presented as isopotential contour plots or as isometric projections, although plots of both potential versus frequency and potential versus cumulative frequency have been used in interpretation.

The interpretation of half-cell potentials with respect to corrosion is also covered by ASTM C876-87 (1987f). Potential mapping can provide information on the extent of corrosion and its probability but rates of corrosion cannot be inferred from absolute values of potential. In general, where corrosion is slow, potential differences rarely exceed 100 mV (ACI, 1985), whereas in reinforced concrete undergoing significant corrosion, potential differences over 200 mV are common. What is important, however, is the gradient between adjacent values because a high potential difference indicates a significant current flow between the test locations. It is important to break out and examine the condition of the reinforcement when undertaking half-cell potential surveys in order to calibrate the potentials measured with the visual condition of the reinforcement.

Fig. 3.7 Use of half-cell potential data logger on bridge column.

Resistivity testing of concrete

The corrosion of reinforcing steel in concrete is an electromechanical process in which anodic and cathodic sites develop on the steel and there is therefore a potential difference between them. As the anode corrodes, a pattern of current flow develops within the concrete, the magnitude of which is directly related to the rate of corrosion at the anode. Theoretically, therefore, one method of determining the probable rate of corrosion of steel in concrete is to measure the conductivity, or inversely the resistivity, of the electrolyte (concrete) as a means of assessing the magnitude of any current flow between anode and cathode.

The method normally used to measure the resistivity of the cover concrete is the four-probe Wenner method (Wenner, 1915) in which four contact probes are inserted into the concrete to a depth of 6 mm with a spacing of 50 mm between the probes. An alternating current is passed between the outer electrodes. The resultant potential difference between the inner electrodes is measured. In practice the resistivity r is calculated from the expression

$$r = \frac{2aE}{I}$$

where r is the resistivity (ohm cm), a is the spacing of the probes (cm), E is the potential drop between the inner electrodes (mV) and I is the current flow between the outer electrodes (mA). The resistivity of the concrete is related principally to its moisture content but is also influenced by its chloride content.

When making resistivity measurements on concrete it is important that the probe spacing should be less than the cover to the steel, to minimize the influence of the relatively high conductance steel. It may therefore be appropriate to locate resistivity measurements diagonally across the reinforcing grid. Problems can be experienced when high resistance surface layers on the structure produce an unduly high contact resistance with the test probes. Aqueous graphite contact fluid can be used to reduce this effect, or it can be overcome by drilling shallow holes to insert the probes in the surface of the concrete.

Values of resistivity are therefore inversely proportional to the conductivity of the concrete to support a certain rate of corrosion. A correlation of resistivity against probability of corrosion is given by Vassie (1980) as follows:

Resistivity greater than 12 000 ohm cm   corrosion unlikely
Resistivity 12 000–5000 ohm cm   corrosion probable
Resistivity less than 5000 ohm cm   corrosion certain if other factors, i.e. half-cell potential, also favourable

Like half-cell potential mapping, resistivity measurements, particularly using surface contact probes, can be a rapid scan technique that provides useful information on the likelihood of the corrosion of steel in concrete. It should be borne in mind, however, that resistivity measurements generally relate to the concrete cover to the steel and not to the resistance of the concrete immediately next to the steel where corrosion is occurring. The method is therefore only indicative and other factors, such as oxygen availability at the cathode or anode–cathode area ratios, may also influence the corrosion rate of the reinforcement. Resistivity testing is normally used only in conjunction with detailed half-cell potential mapping.

3.3.5 Laboratory testing and sample analysis

The laboratory work normally undertaken on samples removed from structures can be subdivided into five types as follows:

1. visual examination of cores, lump samples, pieces of steel etc.;
2. physical testing of the concrete to determine quality;
3. chemical analysis of the concrete to determine composition and the level of chemical attack or deterioration;
4. petrographic analysis;
5. other, e.g. accelerated expansion testing for alkali–silica reaction, abrasion resistance etc.

The types of sample obtained will normally consist of a mixture of concrete cores, powder-drilled samples, lump samples and special samples (e.g. soil and groundwater samples, samples showing efflorescence and exudation etc.) Not all samples are suitable for all laboratory test purposes. If petrographic analysis is required, cored or sawn samples are preferred because lump samples are often badly microcracked and are normally obtained from cover concrete only. Similarly, cement content determinations on powder-drilled samples are to be avoided as powder samples are generally unrepresentative of a unit of concrete.

The number of samples required depends on the type of structure being investigated, its complexity and the reasons for the investigation. If a uniform element of concrete, e.g. a bridge deck slab or column, is being examined it is normal to take at least three samples for testing and to examine at least three areas varying in visual condition from poor to sound. Information from the *in situ* testing programme should be consulted when taking samples. For instance, it would be advisable to take samples for depth of carbonation testing mainly in areas where the cover to steel is low or where the quality of the concrete, from visual inspection or Schmidt hammer testing, for example, is suspect.

Similarly, the locations for powder-drilled samples should ideally be selected after a half-cell potential survey has been conducted so that the areas where corrosion is most likely to be occurring are sampled. Almost inevitably, however, limitations of access, time and economics will be important factors in the selection of sampling locations and numbers of samples.

Guidance on taking core and powder-drilled samples is provided in the literature (BRE, 1986; Concrete Society, 1987; Cement and Concrete Association, 1988) which should be consulted before specifying any testing programme.

On receipt from site all core and lump or sawn samples should be visually examined, described and sketched or photographed before deciding on test procedures and the locations of subsamples for testing.

3.3.6 Visual examination of concrete

An appropriate method for the visual examination of concrete is given in ASTM C856-83 (1987e) for unaided visual examination and low power stereomicroscopy. Guidance on typical defects in concrete which should be noted are given in several references (Dolar Mantuani, 1983; Concrete Society, 1987). The value of the visual inspection of concrete is largely dependent on the experience and expertise of the laboratory personnel carrying out the work. Features which should be examined include the following:

1. The type of coarse aggregate (e.g. gravel, crushed or partially crushed), together with the lithological type(s), grading, porosity and evidence of segregation;
2. the type of fine aggregate (e.g. natural sand, crushed fines etc.) together with the possible lithologies and grading or distribution of the fines;
3. the cement paste — its colour, surface lustre and distribution; whether there is any evidence of microcracking and whether cracks are empty or filled, whether cracks run through or around the aggregate and whether there is any evidence of bleeding or high water–cement ratio porous paste; persistent damp patches;
4. voidage and embedded items, i.e. steel;
5. evidence of major cracks or fractures and any deposits in cracks or voids.

In addition to the visual and stereomicroscopic examination of concrete two further techniques are frequently used in the investigation of concrete deterioration;

1. determination of the air void content and spacing factor of air-entrained concrete (ASTM, 1987b);
2. investigation of the extent of microcracking in concrete by fluorescent dye impregnation or by ultraviolet dye penetration testing.

3.3.7 Physical testing of concrete to determine quality

Physical tests on concrete can be divided into three basic types:

1. testing to determine strength (compressive/tensile);
2. determination of density and excess voidage;
3. testing to assess pore structure (absorption/permeability).

Strength of concrete

The determination of the compressive strength of concrete cores is covered by BS 1881: Part 120 (BSI, 1983b) and in other publications by the Concrete Society (1987) and ASTM (1987a).

   Core samples are normally taken for one or more of the following reasons (Concrete Society, 1987);

1. to estimate the strength of concrete supplied to the structure (potential strength);

2. to determine the srength of the concrete for structural analysis (*in situ* strength);
3. to investigate any deterioration in the structure due to chemical attack, fire etc.

When specifying a core testing programme it is important to decide which of these is the primary reason for taking cores as this will influence both the number of cores required and the locations where the cores should be taken from for testing.

The interpretation of the results of core testing requires an understanding of the factors influencing the strength of concrete in structures. Guidance is given in BS 6089 (BSI, 1981) and in Concrete Society *Technical Report 11* (1987).

It should be borne in mind that core sampling is not always the most appropriate method for the investigation of deterioration in a structure and a number of *in situ* non-destructive tests, e.g. ultrasonic pulse velocity testing, or near to surface tests, such as the BRE internal fracture test (BRE, 1980), may be both more appropriate and more informative. Even where the actual strength of the concrete is in doubt, e.g. in a structure suffering from deterioration due to alkali–silica reaction, the compressive strength of the concrete may be little affected, whereas the ratio of compressive to tensile strength may show a significant change. Similarly where a concrete structure may have been damaged by fire, near to surface tests to assess strength (e.g. the Windsor probe or internal fracture test) or pulse velocity testing may be more beneficial and can be used to assess the extent of any strength loss.

Density and excess voidage

The determination of the density of hardened concrete is covered by BS 1881: Part 114 (BSI, 1983a) which gives methods for the determination of as-received density, oven-dried density and saturated density. A method of determining excess voidage from density measurements on cubes and cores is given in Concrete Society Technical Report 11 (1987).

Testing to assess pore structure

The microstructure of concrete, its porosity and its permeability are the most important factors controlling the durability of concrete where deterioration is due to external agencies, i.e. chloride or carbonation penetration, frost attack or leaching. A Concrete Society report (1988) contains a review and details of techniques for the measurement of concrete permeability and discusses the use of permeability in specifications. However, permeability testing is not generally specified in compliance testing during construction or during repair work because the method(s) are considered to be mainly research based.

However, in the laboratory a number of absorption and permeablity test methods are available to assess the quality of concrete. These include the following:

1. absorption testing (BSI, 1983c);
2. capillary rise testing (Ho and Lewis, 1984);
3. pressure-or concentration-induced flow for gases and liquids (Concrete Society, 1988).

Absorption testing of concrete The method given in BS 1881: Part 122 (BSI, 1983c) involves testing three 75 mm diameter cores which are oven dried for 72 hours and then immersed in water for 30 minutes. The method includes correction factors for 75 mm cores varying in length from 30 to 150 mm. Values for 30 minute absorption (BS 1881: Part 122) of good, average and poor quality concrete are given by Levitt (1984) and others (Concrete Society, 1988) as follows:

Low absorption concrete	3%
Average absorption concrete	3–5%
High absorption concrete	5%

A method for the determination of the absorption of hardened concrete is also given in ASTM C642-82 (1987d).

Capillary rise testing The capillary pore network developed in hardened concrete allows water to be absorbed by capillary attraction. A measure of the rate of absorption of concrete is therefore an indicator of the pore structure of hardened concrete (Fig. 3.8). A number of different test methods are quoted in the literature (Concrete Society, 1988; Lawrence, 1981; Ho and Lewis, 1984) although little guidance is given on typical values for absorption by capillary rise. A value of 10–20 mm absorption after four hours is quoted by the Concrete Society as an acceptable value for concrete.

The test technique is particularly informative in the investigation of frost damage to concrete road slabs where scaling of the surface is due to excessive bleeding of the concrete that produces a weak porous surface layer with a high water–cement ratio.

Capillary rise testing of the outer 50 mm layer of concrete can be used to determine whether the concrete is of uniform porosity and will indicate the presence of more absorptive concrete at the surface of the slabs. The test has also been used to measure the effectiveness of surface hydrophobic sealers such as silanes and waterproofing admixtures/curing systems (Payne and Dransfield, 1985).

Pressure- or concentration-induced flow for gases and liquids A considerable number of test techniques have been developed to determine the coefficient

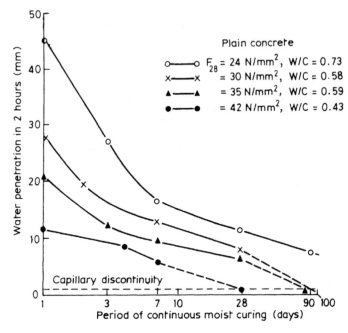

Fig. 3.8 Effect of curing on water penetration by capillary rise for different concrete mixes (after Ho and Lewis, 1984).

of permeability of hardened concrete to liquids and gases. Because of the differences in sample conditioning and test techniques used, absolute values of permeability found by different test methods may vary. Concrete Society Technical Report 31 (1988) quotes the following values for low, average and high permeability concretes:

	Low	*Average*	*High*
Permeability coefficient, water(m/s)	$<10^{-12}$	10^{-12}–10^{-10}	$>10^{-10}$
Permeability coefficient, gas (m/s)	$<5\times10^{-14}$	5×10^{-14}–5×10^{-12}	$>5\times10^{-12}$

3.3.8 Chemical analysis of concrete

The chemical analysis of hardened concrete is covered by British Standards (BSI, 1988a) and is reviewed in a Concrete Society report (1989) entitled 'The analysis of hardened concrete' and in a publication by the Transport and Road Research Laboratory (Figg, 1986).

BRE *Information Paper IP21/86* (BRE, 1986) gives information on taking samples of concrete for both chloride and cement content analysis on site, and on the determination of chloride on site using rapid methods such as the 'hach' chloride test kit and the 'Quantab' chloride test strip.

BS 1881: Part 124 contains methods for the determination of the following:

1. cement and aggregate content (siliceous and calcareous);
2. aggregate grading (coarse/fine);
3. original water content;
4. type of cement (ordinary Portland, sulphate-resisting Portland, low heat, Portland blastfurnace);
5. chloride content of concrete (total acid soluble);
6. sulphate content of concrete (total acid soluble);
7. alkalis as potassium and sodium oxide (fines-rich fraction).

From the point of view of investigations on buildings and structures the following information on sampling, analytical accuracy and specifications for laboratory testing should be borne in mind.

Sampling

The requirements for sampling vary from structure to structure. It is important in cases of dispute that the location and method of testing a sample should be agreed between all parties, as well as the quantity of material which the sample is deemed to represent. BS 1881: Part 124 (BSI, 1988a) recommends the following:

1. The minimum linear dimension of the sample should be at least five times that of the maximum dimension of the largest aggregate particle, i.e. for a 20 mm aggregate concrete the 'minimum' volume should be a 100 mm cube.
2. The sample should consist of a single piece of unfractured concrete if water–cement ratio is to be determined.
3. The minimum mass of any sample should be 1 kg, or 2 kg if the original water content is to be determined.
4. Where information on a batch of concrete is required, at least two and preferably four independent samples should be analysed.
5. Representative samples of the cement and aggregate used, if available, will increase analytical precision.
6. Samples should be clearly labelled to include date, sample location (grid reference, height in the wall etc.) and method of sampling, and should be sealed in a heavy duty polythene bag.
7. Where samples are taken for chloride content testing by powder-drilled sampling, several holes should be drilled to give a minimum 30 g sample.

8. Where samples are taken for chloride content analysis in a group of building components, single powder samples should be taken from 10% of the components. Where the number of components is less than 30 at least three should be sampled (BRE, 1986).

Analytical accuracy: cement content determination and original water–cement ratio of concrete

It is important to bear in mind the accuracy of analytical methods when specifying chemical analysis, particularly in compliance disputes. Guidance on analytical precision is given in Concrete Society (1989).

1. Where sampling and testing variability are included the 95% confidence limits for tests on four independent samples from one batch of concrete are $\pm 40 \, kg/m^3$.
2. The determination of the original water–cement ratio on undamaged uncarbonated concrete with cement content in the range 200–500 kg/m^3 and W/C ratio in the range 0.4–0.8, where aggregates are essentially insoluble in acid, is likely to be within 0.1 of the actual W/C ratio.

Specification of laboratory tests

When specifying laboratory tests it is important to be clear and concise with regard to the reasons why the tests are required and to provide as much information as possible on the samples taken. The analyst undertaking the work should be able to provide guidelines on the appropriate sampling and testing procedures and requirements and should be consulted before a sampling programme is embarked on site. Reinforced concrete is a heterogeneous material and obtaining a representative sample of concrete is fraught with difficulties. The cement content of concrete in walls or columns, for example, can vary by 100 kg/m^3 from top to bottom for a nominally single batch of concrete owing to segregation (Concrete Society, 1989).

3.3.9 Petrographic examination of concrete

The petrographic examination of hardened concrete is covered by ASTM C856-83 (1987e), 'Standard practice for petrographic examination of hardened concrete', and is reviewed in a number of other publications and papers (Power and Hammersley, 1978; Dolar Mantuani, 1983; Baker and Lattey, 1985; Concrete Society, 1987). Specific applications of petrographic techniques would include the following:

1. identification of mix constituents, i.e. coarse and fine aggregate type,

cement type and presence and/or quantity of any cement replacement materials;
2. examination of concrete quality;
3. quantitative determination of mix proportions by micrometric (point counting) methods;
4. diagnosis of deterioration of concrete, e.g. sulphate attack, frost damage, alkali–silica reaction, moisture-sensitive shrinkable aggregates etc.

A number of differing analytical methods are loosely grouped under the heading of petrographic techniques. These include thin and polished section microscopic analysis, electron microscopy and microprobe analysis, X-ray diffraction and infrared spectroscopy. The majority of the test techniques require core samples or sawn lump samples.

Petrographic analysis is concerned mainly with the mineralogy and microstructure of the concrete as distinct from its chemical composition and can therefore provide information on aggregates, cements and concrete which cannot be obtained by normal chemical methods. Information on concrete mix constituents obtained by petrographic analysis can include the following.

Aggregates: coarse and fine

The rock types used as aggregate for concrete can be readily identified by an experienced petrographer familiar with the mineralogical characteristics of the common rocks. Aggregate classification is normally based on the broad rock type descriptions listed in BS 812 (1984). However, it is normally possible by thin section analysis to give much more detailed information on the exact aggregate lithologies and their potential durability.

A full petrographic examination of an aggregate will provide information on the presence of clay and micaceous materials, iron pyrites, soluble salts as coatings or encrustations, organic contaminants, soft or porous particles etc. In addition the petrographer may give guidance on the presence of potentially alkali-reactive rock types. Techniques such as X-ray diffraction and infrared analysis can be used for positive identification of rock or mineral types present. X-ray diffraction, chemical analysis and infrared spectroscopy may also be valuable where, for example, pyrites or swelling clays are suspected in an aggregate used in concrete (Fig. 3.9).

Concrete quality

The petrographic examination of hardened concrete cores can provide information on the *in situ* quality of concrete. Both the relative proportions of cement, coarse and fine aggregate and entrained air and their even

Fig. 3.9 Rust staining due to pyrites on *in-situ* concrete.

distribution can be examined. A semi-quantitative estimate of the relative volumetric proportions of aggregate, cement paste and air can be made by simple visual examination. A more accurate determination may be made by point counting or linear traverse methods. The procedure is described in more detail in ASTM C457-82A (1987b) covering the determination of entrained air in concrete.

Evidence of segregation and bleeding of concrete due to an excessive water content or poor mix design is best examined by means of polished slices of concrete viewed using inclined illumination. This tends to highlight voidage as shadows. Evidence for bleeding, in the form of voids under coarse aggregate particles caused by trapped mix water or bleed channels, can be readily observed.

If thin sections of the concrete are manufactured then a coloured epoxy impregnating resin can be used both to hold the concrete together and to highlight voidage or cracking within the concrete.

Cement paste

The three major constituents of an uncarbonated Portland cement paste can be observed in thin section: amorphous hydration products, portlandite ($CaOH_2$) crystals, and remaining unhydrated cement clinker particles. Observations on the size, abundance and distribution of portlandite and unhydrated cement clinker particles can allow the petrographer to estimate both the water–cement ratio and the degree of hydration of the cement. The size and abundance of portlandite is in part dependent on the water–cement ratio of the concrete. Large calcium hydroxide crystals dispersed through the paste are indicative of a high water–cement ratio whereas in concrete with a low water–cement ratio the calcium hydroxide crystals tend to be submicroscopic.

The size and abundance of remaining unhydrated cement particles will depend on the cement content, the degree of hydration and the fineness of the original cement. In a well-hydrated concrete where the original cement is particularly fine and hydration is advanced, unhydrated cement clinker particles may not be readily visible. At the opposite extreme, an abundance of these particles may indicate that the concrete was inadequately cured.

Cement type and cement replacement materials

Petrography can be used to distinguish ordinary Portland cement, sulphate resisting Portland cement and HAC. The method involves the preparation and examination of polished sections by reflected light microscopy.

Etching the polished section with potassium hydroxide solution allows the relative proportions of the calcium aluminate (C_3A) and ferrite phases in hydrated cement clinker particles to be determined. Sulphate-resisting cement contains relatively little tricalcium aluminate and this is the constituent which is susceptible to sulphate attack (see section 2.2.3).

The presence or absence of cement replacement materials may also be confirmed by thin section examination. Replacement materials which are readily identifiable include pulverized fuel ash and blastfurnace slag.

Diagnosis of deterioration

The petrographic examination of concrete cores, polished slices and thin sections is an invaluable tool in the identification of reasons for failure in concrete structures.

Microcracking The pattern of cracking can be important in the diagnosis of deterioration. For example, cracking parallel or subparallel to the surface of the concrete and within a few millimetres of the surface is often characteristic of freeze–thaw action, whereas cracking perpendicular to the surface, extending from the surface inwards, is often characteristic of excessive shrinkage. Alternatively, cracking associated with a particular aggregate type could indicate that the aggregate is unsound, having undergone alkali–silica reaction or shrinkage.

The mineral ettringite is often found during the petrographic examination of hardened concrete and if present in sufficient quantity, confirmed by X-ray diffraction or chemical analysis, may indicate sulphate attack. In many instances, however, ettringite may be present in thin section without any associated microcracking, indicating chemical deterioration of the paste rather than expansion and cracking.

Deposits Signs of deterioration which are detected during a petrographic examination include the presence of deposits that infill voids and cracks in the concrete, e.g.silica gel in cases of alkali–silica reaction.

Carbonation and lime leaching The observation of an excessive amount of carbonation in a concrete may be important in the diagnosis of the cause of steel corrosion in concrete. Carbonation of the cement paste is obvious in thin section as portlandite crystals in cement paste are converted to minute crystals of calcite ($CaCO_3$). In some instances, excessive carbonation may be only localized, in association with microcracking extending inwards from the surface. In other instances, excessive carbonation may be the result of an initial poor quality porous concrete.

In concrete which has been extensively leached portlandite may be removed from the paste and precipitated as calcium carbonate in cracks or voids or as efflorescence on the surface of the concrete. Attack by acidic waters may remove portlandite entirely from the paste, leaving a soft friable residue of cement and aggregate, or in exposed concrete cladding washed by acidic rain, for example, portlandite may be removed, reducing the alkalinity of the concrete and increasing the risk of reinforcement corrosion.

3.4 Interpretation of condition surveys

The interpretation of the results of condition surveys on concrete structures is, to a large extent, dependent on the experience and expertise of the people undertaking the survey work. A considerable number of different disciplines may be involved, including civil/structural engineering, materials science, concrete and corrosion technology, chemistry and petrography. It is necessary to have at least a reasonable grasp of all of these subjects when interpreting the results of condition surveys on buildings and structures, and frequently advice is needed from specialists in particular fields before conclusions are reached on the condition and future performance of a concrete structure. Ideally a team of different specialists should be involved, although in many cases this is not possible. Advice should certainly be taken when commissioning chemical analysis, for example, and the analyst should be provided with as much information as possible on the reasons why analysis is required and what it is hoped analysis will reveal. This consultation should occur before work on site commences and not after, when all decisions on sample numbers, locations, sizes etc. have already been decided upon and implemented. Although the chemist, petrographer or concrete technologist may not be able to attend on site, he/she can at least be taken through the requirements of the survey, together with any photographs and drawings available, and opinions can be sought on the site works programme and particularly the requirements for laboratory analysis.

It is important at all times to bear in mind the objectives of the survey, as agreed with the client, both to avoid unnecessary work and also to make sure that all the objectives of the inspection programme can be met. Frequently it may be advisable and more efficient to undertake work in stages with an assessment of findings at the end of each stage. The extent and detail of the survey should always be governed by the end use of the information. There is little point in undertaking cement content testing on a structure, for example, if there is no reason to doubt the quality of the concrete or if the analytical result is going to be subject to large errors because of the aggregate lithologies used.

Concrete is an inhomogeneous material and shows a very wide variation in properties depending on the mixes used and how well it is placed around reinforcement, compacted and cured. It is a complex composite material which can exhibit a wide range of properties and consequently can exhibit a wide range of symptoms of distress caused by structural and material inadequacies, early age cracking, poor workmanship and bad design and detailing. When undertaking a condition survey, therefore, perhaps the first thing that should be considered is how the structure was actually built, what governed the design and selection of materials and how the steel/concrete was placed on site. Are there any symptoms of inadequate design or poor workmanship and is the overall appearance of the structure consistent with sound design and construction practice?

The second aspect which should be considered is the loading imposed on the structure and the degree of maintenance/repair work already carried out. Have any joints, for example, been replaced? Has any drainage system present been maintained? Are there any signs of excessive loading/deflection/creep and has any cracking been induced? What are the provisions for accommodating thermal movement and differential settlement and are these satisfactory?

The third aspect is the environment. This should be examined in detail, particularly during periods of wet weather or frost. Are expansion joints functioning and waterproof? Is the structure de-iced and if so where does runoff occur? Are there any signs of ponding or poor drainage inhibiting runoff? On a bridge structure, for example, which elements are exposed to spray from the carriageways or runoff from joints? Which areas are permanently wet and which areas alternatively wet to dry? Which areas are remote from environmental exposure and can be used as control areas?

Finally, when the results of the site and laboratory work are complete and conclusions are being drawn it should be remembered that all the test techniques described are complementary and none, on its own, is definitive. A combination of test and inspection techniques is invariably used to form a pattern for the condition of the structure with each result forming a piece of the pattern. Provided the work has been done correctly there should be no anomalous results and the various test and inspection methods used should act together to build up a picture of the construction, condition and current state of the structure.

As was stated in the introduction to this chapter, well-designed, compacted and cured reinforced concrete is a very durable material. If the durability of a structure is now in doubt then something has gone seriously wrong in its design or construction or in our understanding of the effect of the environment on the life of the materials used.

References

ACI (American Concrete Institute) (1968) *Guide for Making a Condition Survey of Concrete in Service* Committee 201, American Concrete Institute (201.IR-68, reaffirmed 1979).

ACI (1981) *ACI Manual of Concrete Inspection*, Committee 311, American Concrete Institute Special Publication 2, 7th edn.

ACI (1985a) *Strength Evaluation of Existing Concrete Bridges*, American Concrete Institute Special Publication 88.

ACI (1985b) *Corrosion of Metals in Concrete*, Committee Report ACI 222R-85, January–February.

ASTM (1983) *Corrosion of Metals in Associaton with Concrete*, ASTM Special Technical Publication 818.

ASTM (1987a) *Standard Method of Obtaining and Testing Drilled Cores and Sawed Beams of Concrete: ASTM C42-84A*, ASTM Annual Book of Standards, vol. 4.02.

ASTM (1987b), *Standard Practice for Microscopical Determination of Air Void Content and Parameters of the Air Voids System in Hardened Concrete: ASTM C457-82A*, ASTM Annual Book of Standards, vol. 4.02.

ASTM (1987c) *Standard Test Method for Pulse Velocity through Concrete: ASTM C597-83*, ASTM Annual Book of Standards, vol. 4.02.

ASTM (1987d), *Standard Test Methods for Specific Gravity, Absorption and Voids in Hardened Concrete: ASTM C642-82*, ASTM Annual Book of Standards, vol. 4.02.

ASTM (1987e) *Standard practice for Petrographic Examination of Hardened Concrete: ASTM C856-83*, ASTM Annual Book of Standards, vol. 4.02.

ASTM (1987f), *Standard Test Method for Half Cell Potentials of Uncoated Reinforcing Steel in Concrete: ASTM C876-87*, ASTM Annual Book of Standards, vol. 4.02.

Baker, A.F. (1986) Potential mapping techniques. *Seminar on Corrosion in Concrete — Monitoring Surveying and Control by Cathodic Protection*, London Press Centre, Paper 3, May.

Baker, A.F. and Lattey, S.E. (1985) The petrographic examination of concrete. *Proceedings of the 2nd International Conference on Structural Faults and Repair*, Engineering Technics Press, Edinburgh, pp. 129–32.

Bate, S.C.C. (1984) *High Alumina Cement Concrete in Existing Building Superstructures*, Building Research Establishment Report, HMSO, London.

Berke, N.S. (1986) Corrosion rate of steel in concrete, *ASTM Standardization News*, March, pp. 57–61.

BRE (Building Research Establishment) (1977) Determination of chloride and cement content in hardened Portland cement concrete. *Information Paper IP 13/77*, BRE, London.

BRE (1980) Internal fracture testing of in situ concrete — a method of assessing compressive strength. *Information Paper IP 22/80*, BRE, London.

BRE (1981) Carbonation of concrete made with dense natural aggregates, *Information Paper IP 6/81*, BRE, London.

BRE (1986) Determination of chloride and cement contents of hardened concrete *Information Paper IP 21/86*, BRE, London.

Bridge Inspection Panel (1984) *Bridge Inspection Guide*, Department of Transport, Scottish Development Department, Welsh Office, Department of the Environment for Northern Ireland, HMSO.

BSI (British Standards Institution) (1970) BS 1881: *Testing Concrete. Part 5: Methods for Testing Hardened Concrete for other than Strength*, BSI, London.

BSI (1981) BS 6089: *Guide to the Assessment of Concrete Strength in Existing Structures*, BSI, London.

BSI (1983a) BS 1881: *Testing Concrete. Part 114: Methods for Determination of Density of Hardened Concrete*, BSI, London.

BSI (1983b) BS 1881: *Testing Concrete. Part 120: Method for the Determination of the Compressive Strength of Concrete Cores*, BSI, London.

BSI (1983c) BS 1881: *Testing Concrete. Part 122: Method for Determination of Water Absorption*, BSI, London.

BSI (1984) BS 812: *Testing Aggregates. Part 102: Methods for Sampling*, BSI, London.

BSI (1985) BS 8110: *Structural Use of Concrete. Part 1: Code of Practice for Design and Construction*, BSI, London.

BSI (1986a) BS 1881: *Testing Concrete. Part 201: Guide to the Use of Non-Destructive Tests for Hardened Concrete*, BSI, London.

BSI (1986b) BS 1881: *Testing Concrete. Part 202: Recommendations for Surface Hardness Testing by Rebound Hammer*, BSI, London.

BSI (1986c) BS 1881: *Testing Concrete. Part 203: Recommendations for Measurement of Velocity of Ultrasonic Pulses in Concrete*, BSI, London.

BSI (1988a) BS 1881: *Testing Concrete. Part 124: Analysis of Hardened Concrete*, BSI, London.

BSI (1988b) BS 1881: *Testing Concrete. Part 204: Recommendations for the Use of Electromagnetic Covermeters*, BSI, London.

BSI (in preparation) BS 1881: *Testing Concrete. Part 207: Recommendations for Near to Surface Tests for Concrete*, BSI, London.

Cantor, T.R. (1984) Review of penetrating radar as applied to the non-destructive testing of concrete. *In-situ Non-Destructive Testing of Concrete*, American Concrete Institute Special Publication 82, pp. 581–602.

Cement and Concrete Association (1977) *Advisory Data Sheet 34, The Ultrasonic*

Pulse Velocity Method of Test for Concrete in Structures, ADS/34, Cement and Concrete Association, Wexham Springs.

Cement and Concrete Association (1988) *The Diagnosis of Alkali–Silica Reaction, Report of a Working Party*, Cement and Concrete Association, Wexham Springs.

Concrete Society (1987) Concrete core testing for strength, report of a working party, *Technical Report 11*, Concrete Society.

Concrete Society (1988) Permeability testing of site concrete, a review of methods and experience, Report of a Working Party, *Technical Report 31*, Concrete Society.

Concrete Society (1989) Analysis of hardened concrete, a guide to test procedures and interpretation of results, Report of a Working Party, *Technical Report 32*, Concrete Society.

Dolar Mantuani Ludmila (1983) *Handbook of Concrete Aggregates, A Petrographic and Technological Evaluation*, Noyes Publications, New Jersey.

Figg, J.W. (1986) Methods to determine chloride concentrations in in-situ concrete. *Transport and Road Research Laboratory Contractors Report 32*, Transport and Road Research Laboratory.

Figg, J.W. and Marsden, A.F. (1985) *Concrete in the Oceans Technical Report 10*, Offshore Technology Report OTH 84/205, HMSO, London.

Hendry, A.W. and Royles, R. (1985) Acoustic emission observations on a stone masonry bridge loaded to failure. *Proceedings of the 2nd International Conference on Structural Faults and Repair*, Engineering Technics Press, Edinburgh, pp. 285–91.

Ho, D.W.S. and Lewis, R.K. (1984) Concrete quality as measured by water sorptivity. *Civil Engineering Transactions*, Institution of Engineers, Australia, pp. 306–13.

Jones, D.S. and Oliver, C.W. (1978) The practical aspects of load testing. *The Structural Engineer*, **56A** (12), 353–6.

Keiller, A.P. (1982) A preliminary investigation of test mehods for the assessment of strength of in-situ concrete. *Technical Report 551*, Cement and Concrete Association, Wexham Springs, September.

Keiller, A.P. (1985) Assessing the strength of in-situ concrete: an investigation into test methods. *Concrete International*, **7** (2), 15–21.

Kunz, J.T. and Eales, J.W. (1985) Remote sensing techniques applied to bridge deck evaluation. American Concrete Institute Special Publication 88–2, pp. 237–59.

Lawrence, C.D. (1981) Durability of concrete: molecular transport processes and test methods. *Technical Report 544*, Cement and Concrete Association, Wexham Springs, July.

Levitt, M. (1984) In-situ permeability of concrete, *Conference on Developments in Testing Concrete for Durability*, September, Concrete Society, London, pp. 59–64.

Manning, D.G. (1985) *Detecting Defects and Deterioration in Highway Structures*, Report 118, National Cooperative Highway Research Programme: Transportation Research Board, Washington, DC.

Manning, D.G. and Bye, D.H. (1983) *Bridge Deck Rehabilitation Manual*, Part 1: *Condition Surveys*, Research and Development Branch, Ontario Ministry of Transportation and Communications.

Manning, D.G. and Holt, F.B. (1980) Detecting delamination in concrete bridge decks. *Concrete International*, November, pp.34–41.

Menzies, J.B. (1978) Load testing of concrete bulding structures. *The Structural Engineer*, **56A** (12), 347–52.

Mkalar, P.F., Walker, R.E., Sullivan, B.R. and Chiarito, V.P. (1984) Acoustic emission behaviour of concrete. *In-situ Non-Destructive Testing of Concrete American*

Concrete Institute Special Publication 82, pp. 619–37.

Montgomery, F.R. and Adams, A. (1985) Early experience with a new concrete permeability apparatus. *Proceedings of the 2nd International Conference on Structural Faults and Repair*, Engineering Technics Press, Edinburgh, pp. 359–63.

Moore, W.M., Swift, G. and Milberger, L.J. (1973) An instrument for detecting delamination in concrete bridge decks, *Highway Research Record* 541, pp. 44–52, Highway Research Board, Washington, DC.

Parrott, L.J. (1987) *A Review of Carbonation in Reinforced Concrete*, Cement and Concrete Association, Wexham Springs, July, 69 pp.

Payne, J.C. and Dransfield, J.M. (1985) The influence of admixtures and curing on permeability, *Conference on Permeability of Concrete and its Control*, December, Concrete Society, London, pp. 89–105.

Power, T.O. and Hammersley, G.P. (1978) Practical concrete petrography. *Concrete*, **12** (8), 27–31.

Pullar-Strecker, P. (1987) *Corrosion Damaged Concrete: Assessment and Repair*, Butterworths, for Construction Industry Research and Information Association, London;

Savage, R.J. (1985) *Proceedings of the 2nd International Conference on Structural Faults and Repair*, Engineering Technics Press, Edinburgh, pp. 255–64.

Stain, R.T. (1982) Integrity testing. *Civil Engineering*, April, pp. 54–9; May, pp. 71–3.

Tomsett, H.N. (1981) Non-destructive testing on in-situ concrete structures. *NDT International*, December, pp. 315–20.

Vassie, P.R. (1980) *A Survey of Site Tests for the Assessment of Corrosion in Reinforced Concrete*, TRRL Report 953, Department of the Environment, Department of Transport.

Wenner, F. (1915) A method of measuring earth resistivity. *Bulletin of the Bureau of Standards*, **12**, 469–78.

Repair materials and techniques   4

L.J. Tabor
(Fosroc)

4.1 Introduction

It is stated in the Scriptures that it was common knowledge that one should repair a torn garment with sound material of similar performance to the original: that a patch made with unshrunk fabric would soon pull away and cause even greater damage. That same common sense is equally applicable to the selection of materials for the repair of damaged concrete structures, but until recent years had not been developed to its logical conclusion – the formulation of 'pre-shrunk' or non-shrink concrete for repairs.

The choice of material and technique is governed in the first place by the scale of the proposed repair. If a complete concrete member, or the full cross-section of a substantial proportion of a member, is to be replaced, this may be regarded as reconstruction rather than patch repair. In this case it would usually be appropriate to use the original design mix and normal concreting techniques. Nevertheless, a faint shrinkage crack would occur between the recast section and the rest of the original structure and it is common to find specifications calling for this to be subsequently repaired by epoxy resin injection. However, in most repair situations much less than the total cross-section of a member requires removal and replacement. The original concrete mix design is not appropriate for this situation because of the greatly reduced cross-section of pour and in many cases a pouring technique is not really appropriate either; the requirement is for a material which can be packed into the repair zone and be retained with little or no support. These restrictions explain the fairly natural choice of sand–cement mortar for the repair of damaged concrete, a choice often made with reservations and resignation to the fact that, unless executed with exceptional skill, such repairs are seldom durable.

Dissatisfaction with this state of affairs led to the search for better materials and in the mid-1960s polyester and epoxy resin-based mortars became available. Their adhesive bonding capability, rapid development of high strength and outstanding chemical resistance offered clear advantages over the indifferent bond and rapid carbonation experienced with simple cementitious mortar, sufficient to justify their considerably higher cost. Whilst these speciality compositions continue to give excellent service when used in appropriate situations, their indiscriminate substitution soon led to other problems. The rapid curing reactions of these chemicals were often accompanied by a considerable rise in temperature and in some cases substantial shrinkage of the resin binder. When coupled with the subsequent thermal contraction this could induce severe stresses in the parent concrete adjacent to the bond line. A substantial difference between the coefficients of thermal expansion of the resin mortars and their concrete substrates could also exacerbate this situation and give rise to further distress under conditions of thermal cycling. Ignorance of these precepts and the demand for higher strength in shorter time at lower cost caused much disillusionment and damage to the reputation of these materials. Whilst further investigation and testing showed that the problems could be prevented by careful formulation and appropriate specification, these unfortunate experiences spurred interest in the reappraisal of cementitious mortars to overcome their shortcomings by similar careful formulation. The outcome has been a swing back to cementitious repair materials, now greatly sophisticated, but overcoming all the deficiencies of the traditional material. Initially, all the effort went into the development of mortar formulations, since concrete repair practice had largely revolved around hand-applied patch repairs.

The spraying of mortars and concretes for larger scale repair work had a more successful history than mortar-based patch repairs, but the many failures encountered showed that there was still a need for greater understanding and development of the materials used. Much of the recent development in the technology of patching mortars can be directly translated to the spraying technique, and some of these products have proved suitable for spraying without further modification.

The latest development in concrete repairs is the pouring of concrete behind a fixed shutter and appears to have brought technology full circle, but there are some dramatic differences. This is concrete designed to be poured into thin (50–100 mm), large area, vertical or overhead patches and to be completely self-compacting around congested reinforcement. The use of pumpable self-compacting concretes for large pours in new construction has become widely accepted and their flow properties are measured on a flow table (BSI, 1984). These new repair materials, however, are so fluid that they run completely off the table, and so a flow trough of the type used for grouts is employed to monitor their performance.

It is inevitable that the development of new materials has required new test methods, but more than that, the quest for durability in concrete has led to intense efforts to understand the complex mechanisms of deterioration and to identify and quantify those characteristics which influence durability. In particular, the development of methods of measuring the various diffusion parameters has been of great value in helping the material formulators to appraise the course of their developments.

The development of good diffusion barrier properties in cementitious repair compositions is now recognized as being of equal importance to their strength and bond. Whilst the resistance of these compositions to the diffusion of chlorides and carbon dioxide has been improved one or two orders of magnitude beyond that of good concrete, the need to upgrade the diffusion resistance of adjacent concrete to give it comparable durability with the repaired areas has been met by parallel developments in the field of surface coatings.

In this chapter we also consider the techniques of repair, the breaking out of concrete, preparation of the steel, mixing, application and finishing of the replacement mortar or concrete, but first, before the materials are considered in more detail, it is important to consider their origins.

4.2 Materials for concrete repair

4.2.1 Sources

The sourcing of materials for normal construction quality concrete is comparatively simple. Portland cements to BS12 (BSI, 1978) are readily available from some 26 plants in the UK and many others around the world. Although there are differences due to raw material sources and process variations, all are deemed suitable for producing good quality concrete. Aggregates show greater variation but their limitations and the mix design procedures for obtaining the best results from them are well understood. Their production methods are quite unsophisticated. The only other essential ingredient is potable water. The understanding of these materials and the production of good concrete from them is part of the stock-in-trade of every civil engineer and builder.

The contrast with the sourcing of repair materials could not be greater. Synthetic resins, both epoxy and polyester types, are produced from petrochemicals and the diversity of the basic resins is multiplied exponentially by the wide range of modifiers blended with them to obtain particular properties. There are, as yet, no standards for these or for the newer cementitious repair compounds, which again can be exceedingly complex in composition. This is the field of the materials scientist and the formulating chemist. It soon becomes apparent that this sophistication of repair materials has introduced another actor to the scene, interposed between the basic commodity supplier and the management of the construction (or repair) site: the specialist material formulator.

4.2.2 The specialist formulator

The specialist formular is not, as some imagine, someone who buys bulk materials from chemical suppliers, breaks them down into small packages and resells them at a considerable mark-up. Whilst there is a limited need for such a service, that is really a merchant activity. The specialist formulator employs materials scientists and chemists to take basic materials from the petrochemical, plastics, cement, aggregate etc. industries and to tailor them into the complex materials required on site, taking special care to make their application on site as simple as possible. Thus a resin mortar may contain a basic epoxy resin, modifying diluents, pigments, suspension agents and a trace of solvent; the hardener component may have three or four ingredients reacted together and the filler may be a blend of several prepared sands. A cementitious repair mortar may be just as complex, containing Portland cement, blastfurnace slag, microsilica, a tightly controlled addition of expansive cement, plasticizers, polymer powder and set control additives.

Such sophistication requires extensive laboratory facilities for the development and short-term and long-term testing of the formulations, for control of their production and for provision of technical support to the specialist applicators and the specifiers, consultants etc. for whom these materials may present a new discipline to be grappled with. The annual budget for one UK company's research and development technical support to their worldwide effort is some £3 million. This includes funding of university research programmes, active support of professional and technical bodies such as the Concrete Society and the Construction Industry Research and Information Association (CIRIA) and the presentation of papers at national and international conferences.

The industry has a trade association, FeRFA*, composed of specialist formulators and applicators, which seeks to promote and maintain standards of commercial activity and technical expertise. Through its various committees, codes of practice are established, technical problems resolved and, in conjunction with the British Standards Institution, the Department of Transport etc., standard methods of test and appraisal are evolved. There is now a European equivalent known as EFNARC.†

4.2.3 Replacement materials – the choice

The first consideration is the choice between resin-based and cementitious materials. As has been indicated above, the development of cementitious materials has now reached the stage that they should be the first choice, unless there are other factors which preclude them. Time is often a major consideration, and although fast-setting comentitious materials are available these can normally only be used for floor, road or deck repairs where they can be quickly puddled into place and left to set. Some resin mortars, however,

---

*FeRFA, The Trade Federation of Specialist Contractors and Material Suppliers to the Construction Industry. 1st·Floor, 241 High Street, Aldershot, Hants GU11 1TJ.
†European Federation of National Associations of Specialist Repair Contractors and Material Suppliers to the Construction Industry (same address and secretariat as FeRFA).

have sufficient working life to allow their application by trowel in vertical and overhead situations and still gain strength rapidly after gelation. Systems based mainly or entirely upon lightweight fillers are preferred for overhead work. Dense high strength resin mortars may be preferred where high impact or abrasion resistance is necessary, e.g. for vulnerable arrises at the corners of buildings or abutting movement joints in warehouse floors. Similarly, an overall resin mortar screed may be required where good concrete has been found to have inadequate resistance to a particular abrasive or chemically aggressive environment.

Resin mortars and cementitious systems are discussed in greater detail in the sections which follow.

4.2.4 Resin mortars

The advent of resin mortars into the concrete repair scene has been introduced briefly in section 4.1 and is reviewed in more detail by Tabor (1985). As is often the case with newly developed materials, it took some time to establish their correct niche in the spectrum of construction materials. For a few years they appeared pre-eminent in the concrete repair field but, as was discussed in section 4.1, that place has now been taken over by more sophisticated cementitious formulations. Resin mortars still have a role to play in concrete repairs but it is less all-embracing than was envisaged in those early years. They have some attributes which can never be matched by cementitious materials: resistance to a wide range of aggressive chemicals, the rapid development of high strength and the ability to cure or harden under environmental conditions outside the range of their cementitious counterparts.

The chemical resistance of cementitious mortars and concretes is very limited and so they have been replaced on a large scale by resin compositions in the field of industrial flooring. Where this need has not been anticipated at the design stage, or where a change of use has ensued, the deterioration of concrete by chemical attack may dictate the use of resin mortars for repair work. These mortars are blends of reactive resin binders with graded fillers, usually silica sands and ground mineral powders. The binder may be a polyester resin or an epoxy or modified epoxy type. Other sophisticated resins are used as binders in flooring compositions to meet very specific needs, but seldom find their way into the concrete repair market.

Both polyester and epoxy resin binders are thin syrupy liquids which remain stable for a long period under suitable storage conditions. They harden rapidly by a chemical reaction which is brought about when a second component, usually called the hardener, is added and stirred in.

Polyester resins

Polyester resins have the advantage of simplicity in this respect because all the necessary components of the curing reaction are contained in the liquid resin; the hardener is simply a filler containing a powdered catalyst which initiates the reaction. The precise quantity of catalysed filler is usually not very critical as long as a minimum threshold quantity is added. This gives scope for the filler content to be adjusted to suit the work in hand, so that the product may be employed as a semi-fluid grout, a trowelable mortar or a stiff mix to be rammed into place. This is of course a generalization and the formulator's instructions must be heeded if the shrinkage is to be controlled properly and the appropriate mechanical properties are to be achieved, but it does make these materials tolerant of the rough-and-ready treatment they are likely to receive at the hands of many site operatives.

Polyester resins are capable of very rapid reaction but in the course of this they undergo a high degree of shrinkage. By careful formulation and a sufficient quantity of well-graded fillers the shrinkage can be reduced to an acceptable degree, but it still imposes a limitation upon the volume of material that can be placed at any one time.

Epoxy resins

Epoxy resins harden by an additive chemical reaction, i.e. by the addition of a reactive hardener, not a catalytic one. In this case the hardener is a reactive liquid whose quantity must be measured carefully to match the chemical requirement of the syrupy resin base. An excess of either component can mean that the hardened resin is softened and weakened by the effect of the unreacted liquid. Whilst progress has been made in developing formulations which are tolerant of a small excess of either component, considerably more care needs to be taken to ensure that these two components are used in their correct proportions. Most of the responsibility for this proportioning is taken out of the hands of the site operative and the specialist formulator supplies the products in a package of a size appropriate to typical site needs, the resin base, the hardener and the filler being separately weighed at the factory into their respective containers. It is still incumbent upon the site operative to ensure that the full quantities supplied are thoroughly blended together. The reaction of epoxy resins causes a slight reduction in volume in the liquid stage, but once gelation occurs there is virtually no further shrinkage as a direct result of the reaction although, as with polyesters, if the hardening occurs rapidly there will be a rise in temperature (because the reaction is exothermic) and consequently some thermal contraction on cooling.

Epoxy and polyester mortars

Whilst resin compositions may be used as grouts for filling cracks (section 4.2.6) and bonding-in starter bars etc., the usual requirement in concrete repairs is for mortars. Polyester mortars, as has already been stated, are generally confined to small works, and so are commonly employed in the repair of chipped arrises and localized floor damage etc. They can be quite a boon in the precasting yard and on the construction site where high strength repairs may be required with the minimum of delay.

In a well-formulated epoxy mortar the shrinkage can be as low as 20 microstrain (in fact a suitable method of measuring it reliably has only recently been developed) (Staynes, 1988). As a result, epoxy mortars can be used for much larger repairs than polyesters, although the effect of bulk upon the build-up of exothermic heat must always be borne in mind. Thus, when used at 5 mm thickness, epoxy mortars can be laid over almost infinitely large areas and so are used for heavy duty floor screeds. In the repair role they are usually required in somewhat thicker sections and so the length of run applied at one time may need to be restricted to avoid excessive thermal contraction strain.

Polyester mortars are usually sufficiently resin rich that they 'wet' the concrete without the need for any separate primer, so adding to their simplicity. Epoxy mortars, because of their higher material cost and different application properties, are usually formulated with no excess resin to achieve the necessary wetting. A separate primer is therefore needed. This is usually an unfilled resin/hardener (sometimes containing solvents) and must be brushed on to the prepared concrete substrate to give intimate contact and continuity of the resin matrix between the resin mortar and the host concrete.

Although the cost of resin mortars may be high, it is often a small proportion of the cost of the whole repair exercise. The rapid development of serviceable strength (typically in an hour or two for polyester resins and a day for epoxy resins) may often be sufficient to justify the cost. Where chemical resistance, high strength and high abrasion resistance are required there may be no cheaper alternative.

The strengths attained by these materials are of the order of 50–100 N/mm^2 in compression, but perhaps more impressive are their tensile and flexural strengths, which are respectively about four times and six times those of ordinary Portland cement (OPC) concretes. This gives them superior performance under conditions of severe impact and abrasion and accounts for their use in the repair of floor joint arrises, stair treads and other similarly demanding applications. They are often used for the repair of floors prior to overlaying with a flooring composition of one sort or another, the advantage over a cementitious mortar being the rate of hardening and the absence of moisture to delay the application of the topping.

For more general concrete repairs to beams, columns, walls and soffits where cementitious materials are precluded for whatever reason, it has already been stated that polyester mortars are used for small patches and epoxy mortars for larger ones. Often there is no need for very high strength and advantage can be taken of the availability of lightweight fillers to permit higher build in vertical and overhead situations. Lightweight epoxy mortars with a relative density of around 1.2 can be applied in these difficult situations to thicknesses of 50 mm or much greater in smaller pockets; they achieve a compressive strength of 40 N/mm^2 and still have better impact strength than concrete.

Because of their amorphous nature, resin binders have coefficients of thermal expansion many times greater than that of concrete (typically $65 \times 10^{-6}/°C$ compared with $12 \times 10^{-6}/°C$ for concrete). Although the addition of mineral fillers reduces this differential to about 2.5 there is a risk of differential thermal stresses developing at the bond line if a concrete–resin mortar composite is subject to changes in temperature. Whilst this can cause total disruption of the joint, with careful formulation and intelligent design the effect can be neutralized. The differential stress developed at the interface by a change in temperature is dependent upon the differencce in the thermal coefficients of the two materials and their relative elastic moduli. Whilst a heavily filled resin mortar can be formulated to have a modulus value approaching that of concrete (about 28 GPa), the differential thermal stress can be reduced by using a slightly more flexible resin. Furthermore, the lower modulus resin systems have some capacity to creep under load, allowing differential thermal stresses to decay fairly rapidly. By taking careful account of these factors, formulators have produced epoxy resin mortars which, in the case of flooring systems, can withstand the thermal shock of steam cleaning without impairing their bond to the concrete substrate. Thus the difference in thermal coefficients need not impair the durability of bond when epoxy mortars are formulated for concrete repair work.

ASTM C-884 (ASTM, 1983) puts these considerations to the test and examines the compatibility of a resin mortar topping with a cementitious substrate. By subjecting a composite specimen to thermal cycling, any development of excessive stress at the bond line is readily revealed by the onset of delamination.

The implications of a mismatch in properties between the repair system and the substrate on the structural behaviour of a repaired concrete member is discussed in some detail by Emberson and Mays (1990 a and b). Short-term problems may arise because the repair materials contract during cure relative to the surrounding concrete. During service, incompatibilities in the form of differing elastic moduli may affect the transmission of load through the member. Additionally, creep of the repair material under sustained stress may cause the repair to become less effective with time.

4.2.5 Cementitious repair packages: examining the systems

An outcome of the intensive activity in developing cementitious repair materials has been the appearance on the market of complete 'repair packages' from most of the specialist formulators. These usually include not only a fairly sophisticated repair mortar but also a bond coat or bonding bridge to promote good adhesion between the repair mortar and the concrete substrate, an anti-corrosion primer for the steel reinforcement and an anti-carbonation coating to be applied over the whole structure after repair. Before the latter can be applied, however, it will usually be found that a fairing coat or levelling mortar is required to fill in blow-holes and irregularities in the areas of concrete not needing repair, in order to create a uniform and tightly closed surface upon which to apply the anti-carbonation coating (Fig. 4.1).

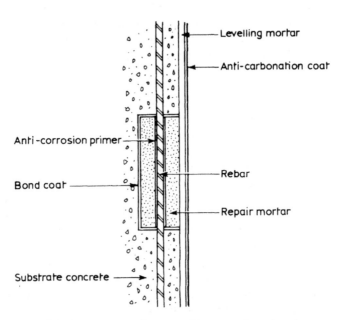

Fig. 4.1 Elements of a cementitious 'repair package'.

The engineer, presented with the proliferation of materials which are now deemed necessary to carry out what he had previously considered a simple straightforward repair, may be forgiven for cynically suspecting that this is just so much gimickry designed to boost the formulator's sales. The ensuing paragraphs will examine each of these product groups in turn and ask whether they are really necessary.

Treatments for rusty steel

The majority of concrete repair contracts are for the reinstatement of reinforced concrete. This usually means that the steel reinforcement has rusted to some extent. It is generally considered necessary to remove the corrosion products from the steel before proceeding with replacement of the concrete. Some claims have been made that this expensive and time-consuming removal of rust can be avoided by the use of chemical rust converters. Phosphoric-acid-based rust converters have been on the market for many years and have apparently found some success in the surface coating industry. They are not appropriate to the concrete repair situation, however, because the phosphate coating they produce is water sensitive and requires protection with a chlorinated rubber coating before concrete may be poured against it. Other rust converters are available based on tannins and these appear to be a more helpful proposition. However, some recent unpublished trials in the UK have indicated that, whilst they give good performance in the repair zone, their use may increase the potential for corrosion immediately beyond the repair.

In considering preparation of the steel reinforcement we have already strayed into the field of corrosion inhibitors. This is a very complex subject in which the leading workers stress the need for more research, investigation and long-term trials. As hinted in the paragraph above, it is important to consider the relationship between the repair zone and the hitherto sound reinforced concrete beyond it.

This is particularly pertinent where resin mortars are used for the repair. There has been much speculation but, to the author's knowledge, no con-clusive research into the effect of encasing part of a reinforcement bar in one of these electrically insulating mortars. Whilst there is no doubt that these materials can give good barrier protection to the steel with as little as 5 mm cover, unless the steel has received an anti-corrosive treatment there is a very real risk that rust originating beyond the repair zone may creep along the bar beneath the epoxy system.

Cementitious coatings When a cementitious repair material is being used it may well be argued that this should provide the ideal alkaline environment for protection of the steel, without any further treatment. However, because hand-placed mortars are much drier than normal concrete at the time of application, it is common practice to apply a cementitious slurry coating to the steel reinforcement to ensure intimate contact of the alkaline protective medium. Some early specifications called for a simple OPC and water slurry to be used, but this proved to be of little practical benefit. Such slurries were found to have very poor application properties, they thickened rapidly (and so were likely to be repeatedly diluted) and soon dried to a dusty powder of unbonded partially hydrated cement. These difficulties were soon over-come by the incorporation of polymers, typically styrene-butadiene rubber (SBR) or acrylic dispersions, which ensured a coherent well-bonded film of cementitious slurry. Excellent though it may appear, this modification of the slurry may raise more questions than it answers. The polymer modification of floor screeds has long been shown to be beneficial in improving their mech-anical performance and abrasion resistance and, as something of a bonus, their chemical resistance. The polymer forms a protective film around the cement particles and gives enhanced protection against acid attack. When the slurry coating for steel reinforcement is modified in a similar way (usually to a somewhat higher polymer–cement ratio) the same polymer 'protection' of the cement could reduce its alkaline benefit to the steel. This was one of the findings of the author's unpublished research a few years ago when polymer–cement slurry coatings were found to be less effective than an unmodified cement slurry in protecting steel during exposure to a severe salt-spray regime. This conclusion related only to slurry coatings on steel and should not be interpreted as an indictment of polymer modification of cementitious mortars, which is discussed later in this section.

Epoxy resin coatings Another approach to steel reinforcement protection adopted by some system formulators is the use of epoxy resin coatings, usually containing an anti-corrosive pigment such as zinc chromate, iron oxide or even ground cement clinker. (The latter is claimed to provide an alkaline environment in which the steel will not rust, although it is difficult to conceive how this can function in a chemically resistant non-ionic epoxy resin medium.) Epoxy resin coatings have a long history of success in the protection of external steel work. They have also been shown to give good protection to steel reinforcement when it is factory coated prior to incorporation in reinforced concrete. Indeed the use of epoxy-coated reinforcement in the construction of concrete bridge decks is mandatory in Canada. Whilst these systems have proved to give good protection when the entire bar is coated, it may be unwise to apply coatings when only a limited length of bar is accessible. A few years ago it was found that, although the coatings provide excellent barrier protection of the steel, if corrosion is initiated beyond the coated area rusting may progress along the steel underneath the epoxy coating.

This is explained by the work of Gledhill and Kinloch (1974) who showed that if water can get to the epoxy–steel interface the bond becomes thermodynamically unstable. Debonding occurs, followed by rusting. This view appears to be endorsed in the findings of a recent Building Research Establishment (BRE) report. When examining 'repaired' reinforced concrete specimens after ten years of weathering in the tidal zone of an exposure station, Cox *et al* (1989) observed: 'Also of significance was the fact that invariably where repairs based on Portland Cement were made, corrosion in the repair region covered most or all of the surface area (of the reinforcement) where a barrier primer to the steel was used. However where no barrier protection was used, the total area showing corrosion was much less.'

Similar conclusions were reached by McCurrich *et al* (1985) in investigations of the effect of accelerated corrosion cycling of concrete prisms incorporating steel reinforcement, part of which had been exposed and treated with one of a range of anti-corrosive systems, with the missing concrete replaced with a cementitious mortar repair. They found it most illuminating to examine not only the steel in the 'repaired' zone but also the undisturbed steel in the adjacent concrete. With shallow cover and the severely corrosive 52 week hot salt-spray/freeze–thaw/drying cycle to which they subjected the specimens, rusting progressed beneath epoxy coatings and polymer-modified cementitious slurries in the 'repaired' zone. When a plain cement slurry was used to pretreat the steel in the 'repaired' zone, this part of the bar was found to be completely free of rust when the specimens were broken open at the end of the accelerated corrosion cycle, but rusting in the undisturbed part of the bar was found to be severe. This suggested that the slurry coating provided a more highly alkaline environment than the mature concrete further along the bar, making the 'repaired' zone cathodic and accounting for its good protection at the expense of the adjacent length of bar, which had thus become a sacrificial anode. The beneficial effect of a sacrificial anode was clearly demonstrated in another specimen where the exposed length of bar was not treated or covered in any way. This section corroded very severely, but gave protection to the adjacent embedded steel which, by virtue of the effect of the alkaline concrete casing, was cathodic with respect to the exposed length. When the specimen was broken open the embedded length was found to be in a clean bright condition, which could only be due to the sacrificial effect of the exposed rusting bar, because this shallow concrete cover had been shown on the earlier specimens to be inadequate protection to the steel under this severely corrosive regime.

Zinc-rich epoxy coatings The rusting of steel is an electrolytic process. It therefore makes sense to eschew barrier coatings, whether alkaline or not, and to adopt an electrolytic or galvanic approach to its prevention. This was the philosophy behind the final systems investigated in the McCurrich trials: a zinc-rich epoxy coating was used. Not only did the zinc-coated 'repair' zone remain rust free, but the encased steel beyond the repair remained in a clean condition. Whereas the strongly alkaline cement slurry primer on other specimens had induced corrosion in the previously good steel beyond the repair (and particularly at the junction between the repair mortar and the host concrete), the zinc-rich coating produced a completely reversed electrolytic balance, keeping the repair zone anodic, so that the untreated areas beyond the repair were in a relatively cathodic condition with respect to the repair. The repair mortar kept the zinc coating in a stable condition so that the 'repaired' and adjacent areas were maintained in a healthy equilibrium throughout the trial. Zinc-rich epoxy coatings complying with BS 4652 (1971) give a dried film containing more than 95% of metallic zinc, so that there is insufficient resin binder to effect any sort of barrier to the steel and electrical isolation does not occur; contact between the metallic particles ensures a conductive path between the repair mortar and the steel.

It has been suggested that zinc-rich coatings should not be used in conjunction with cementitious materials because of the susceptibility of zinc to alkaline attack. This is not the case: there is some slight reaction at the interface but it is not progressive and is in fact beneficial in enhancing the bond between the mortar or concrete and the coated steel.

It is interesting to note that more than one major formulator now offers a zinc-rich coating for steel reinforcement in conjunction with both cementitious and resin mortar systems.

Bonding coats

Nearly all the prepackaged cementitious concrete repair systems on the market contain some form of bonding coat or primer to promote the adhesion of the repair composition to the concrete substrate. The concrete engineer sometimes expresses surprise that this should be considered necessary. He can accept that resin mortars may require a primer to be applied to the concrete because the bond line is an interface between two very different materials, but he is understandably suspicious when it is suggested that a primer is needed to promote the adhesion of fresh concrete or cementitious mortar to an existing hardened concrete substrate. He will have ample experience of the sound bond which can be achieved between new concrete and old when all the rules of good concrete practice are followed. He will also confess, albeit somewhat reluctantly, to have seen cases where good bond was not achieved in spite of all the best endeavours, particularly in the field of flooring, where partial or total debonding of screeds is all too common, and also in concrete repairs, where the shelling-off of sand–cement mortar patches has often been part of the continuing cycle of building maintenance. For some engineers the frequency of these failures has led them to conclude that new concrete has no adhesive properties. This is demonstrably not true, but this dichotomy of view and experience has bedevilled concrete engineers for many years.

Slurry bonding coats Practicing floor-layers found that a good bond was most likely to be achieved by taking steps to minimize shrinkage and ensure good contact between the mortar and the dampened substrate. To minimize shrinkage they needed to use the minimum cement content compatible with strength requirements, keep the water–cement ratio to a minimum by using an 'earth-dry' mix and finally exercise good curing control to minimize water loss. The use of an earth-dry mix made the good contact between mortar and substrate difficult to achieve, and so it was found beneficial to prime the previously dampened substrate with a cement–water slurry, compacting the mortar into this before it dried out.

It was later found that this slurry coat could be given real adhesive properties by replacing about half the water by a natural rubber latex or a synthetic polymer dispersion such as polyvinyl acetate (PVA), SBR or an acrylic latex. Alkali-stable grades were necessary to avoid coagulation of the rubber or synthetic polymer (see later in this section for a discussion on the selection of polymer dispersions).

This type of slurry bonding coat has been proved over many years to give very good insurance against the development of curling in floor screeds. It is equally effective in the concrete repair field, but has been called into question in recent years by failures that have occurred when the bonding coat has been allowed to dry out before application of the repair mortar. When this happened some workers simply applied a second bonding coat and some manufacturers even advocated the use of a double coat as standard practice. One investigation into such failures revealed that the dried bonding coat was not receptive to a further application and that the bond between two such coats could be very poor (Dixon and Sunley, 1983). This particular work was done with an SBR dispersion and it is unfortunate that the workers came to the conclusion that SBR bond coats were unsatisfactory, instead of condemning the practice of double bond coats. They further questioned the value of bond coats altogether, because their slant shear bond strength tests (BSI, 1984, BS 6319) showed that the purely cementitious bond achieved by the application of a sand–cement mortar directly on to the prepared susbtrate was at least as good as that obtained when a polymer–cement slurry bond coat was used.

The author then carried out corresponding trials and found that the serious reduction in bond strength with 'double bond coats' also occurred with acrylic dispersions; it was not a peculiarity of SBR dispersions.

Development of bond strength When laboratory results conflict with the experience of years of site practice it is wise to question whether the test used was really appropriate. Whilst the slant shear bond test is a valuable tool in the assessment of materials in various aspects of concrete repair, it is perhaps not the most relevant to this situation and should not be used in isolation. The slant shear bond test evaluates the performance of the fully developed bond when working stresses are applied to the concrete, but when the repair patch is first applied or the floor screed is laid, the stress most likely to develop at the interface before full service conditions are imposed is a tensile one due to early shrinkage strain (albeit imperceptible) tending to lift the screed or patch at the edges. Thus the early development of good tensile bond strength is a prerequisite for achieving a good long-term bond.

If 1.5 N/mm^2 is taken as a satisfactory tensile bond strength, it will be seen from Fig. 4.2 that this is attainable in a purely cementitious system, i.e. simply by wetting the prepared substrate and applying fresh concrete or mortar (if compaction is good, no further improvement is obtained by the use of a cement slurry primer). It can also be seen that this level of bond strength is only attained after some seven days at 20°C (two to three weeks at typical site temperatures). It is the author's contention that if the bond is subject to stress during that period (the most likely cause being slight drying

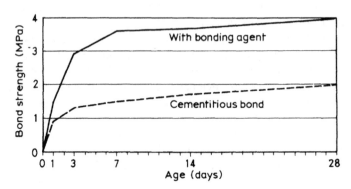

Fig. 4.2 Development of tensile bond with time.

shrinkage) the bond will be ruptured and will not re-form. This is the key
to the dichotomy of conflicting experience referred to earlier.

Further reference to Fig 4.2 shows that when a bonding aid is used the rate
of bond strength development is much more rapid and the critical 1.5 N/mm^2
can be achieved in the first 24 hours, after which the bond is no longer vulnerable. This then explains the very real benefit of bonding aids proven in the field.

Application of bonding coats Since the reapplication of a polymer–cement bond
coat is not to be recommended, it follows that great care must be taken to apply
this type of bond coat just in advance of the mortar or concrete application,
but as the rate of drying of any water-based material can be so variable and
unpredictable there is always a risk that premature drying may occur. If it
does, the only recourse is to wet-scrub back to clean concrete and start again.

One manufacturer of prepackaged repair compositions claims to have
circumvented this problem by using a polymer dispersion alone without any
cement. Although the drying of these aqueous polymer dispersions on their
own must be equally variable and unpredictable, the films which they form
are not immediately water resistant: in fact they may take several hours or
even days to develop their full wet-scrub resistance. Some can be shown to
be reactivated up to 24 hours later by the application of a wet cementitious
mortar and still develop a good strong bond. This may be quite acceptable
as a working practice when the repair is in a horizontal deck situation, but
in vertical and overhead situations the bond coat is required to afford some
'grab' to the repair mortar to help hold it in position until it has hardened.
This can be achieved by the application of a second coat of those systems
that have been shown to have this reactivation capability.

The benefits to be obtained from the stickiness or 'grab' of various bond
coats have been measured in a simple test developed by Judge and Lambe
(1987). This parameter, together with the wet cohesive strength and density

of the mortar, determines the depth of build which can be achieved in one application. These are matters which warrant yet further study.

Prior to the application of any aqueous bond coat, whether slurry or straight dispersion, it is important that the concrete substrate be thoroughly saturated. This is best achieved by spraying lightly with water either continuously or at frequent intervals until no more absorption takes place and then ensuring that any surface water is removed; the bond coat is applied when the surface is just beginning to dry.

There are situations where the saturating of the substrate is undesirable (section 4.3.4.) and these bonding coats cannot be used. The alternative is to use an epoxy resin bond coat. These are two-part systems requiring the correct proportioning of resin base and hardener (usually supplied in pre-weighed packs) which must be mixed thoroughly and then applied to the prepared concrete substrate. The fresh cementitious mortar or concrete should then be applied whilst the epoxy resin is still in a tacky condition. Suitably formulated epoxy bond coats usually have an 'open time', during which the cementitious material can be placed, of up to 24 hours. In fact, it has been the author's observation that the most successful systems allow the concrete to harden before the epoxy. The suitability of a formulation should in any case be checked by carrying out a slant shear bond test. Some formulations have been put on the market in the naive assumption that any two-part epoxy formulation will work. This is demonstrably not so as neither the very water-resistant nor the very water-tolerant types of hardener will have the desired effect. This was clearly illustrated some years ago when the author was measuring the variation of bonding capacity with open time using a number of commercial formulations. One of these, which was too water tolerant, gave poor results when concrete was applied to the fresh resin coating, but when the resin coat was left open for one or two hours a much higher bond strength was obtained. The partially reacted resin, although still very tacky, had become less water tolerant and had achieved the optimum hydrophobic–hydrophilic balance. Better formulated materials maintained this balance throughout their open time.

Resin bond coats can also be of advantage in situations where the substrate concrete is likely to remain permanently saturated, when a polymer-dispersion based bond coat might never develop full strength.

Cementitious mortars

At the time of writing the vast majority of concrete repairs are carried out by the hand application of cementitious mortars. These are almost invariably prepared from sophisticated factory prepacked powder blends, usually modified by the addition of a polymer emulsion or dispersion. Before looking at the factors accounting for their success, it is worth considering why the earlier repairs using simple sand–cement mortars led so often to early failure.

To produce a simple sand–cement mortar with good workability under the trowel required a cement:aggregate content of the order 1:2.5 to 1:3 and a fairly high water content. Thus a high level of shrinkage was inevitable and this accounts for the cracking and debonding which occurred so frequently. It also gave a high degree of porosity which led to rapid carbonation and susceptibility to frost damage. Invariably these materials were batched on site with no control and so there was great variabilty.

The exceptions to this saga of failure were the very few specialist contractors who used an earth-dry mix which they firmly rammed into placed and then cured with great care. The other exceptions were works carried out by spray application, which gave the opportunity to achieve good compaction throughout the depth of the repair whilst maintaining a low water–cement ratio (section 4.3.8).

The first step in the improvement of cementitious repair mortars was the introduction of polymer emulsions. More correctly described as dispersions, these milky-white aqueous liquids usually contain 45%–50% suspended polymer solids. Many types of polymer dispersion are suitable for this work but there is a great deal of misunderstanding about their selection. Various arguments, fuelled by commercial bias, have led to the issue of a number of dogmatic statements and demands from ill-informed specifiers. Whilst straight vinyl acetate dispersions are unsuitable for this purpose (because they can hydrolyse under damp conditions) and are no longer offered by any reputable formulator, various vinyl acetate copolymers, particularly those with Veova, are eminently suitable, as are many SBRs, acrylics and styrene-acrylics. It is most unwise of the specifier to insist upon a dispersion from one of these groups to the exclusion of others because there are no broad distinctive differences in performance between the groups, but within each group there can be wide variations in performance depending on the proportions and types of monomers employed in their manufacture, the amount and type of dispersion agent, surfactant, stabilizer, anti-oxidant etc. The selection of a specific polymer dispersion from this bewildering range can only be undertaken by an experienced formulating chemist having due regard for the parameters to be met and the other materials to be selected to achieve these objectives. The only sensible demands which the specifier can make are those based upon performance of the total formulation, not its composition. It is to be hoped that the issue of the revision of the Concrete Society's *Technical Report 9* 'Polymer concrete' (1975) will reassure engineers and specifiers on this matter.

The incorporation of polymer dispersions into site-batched sand–cement mortars improved their bond and increased their tensile strength and so reduced the tendency for cracking and debonding. It also acted as a plasticizer and reduced the water–cement ratio, but further sophistication of the mix was required, together with closer control of site activities, before these problems could be eliminated and consistent performance achieved. The next step then was to bring the batching of the ingredients under control by factory weighing and blending, which made it feasible to introduce other refinements: admixtures to reduce the water requirement, improve workability and reduce the water absorption of the cured product; a more complex blend of sands and powdered mineral fillers to give better compaction, lower permeability and, in many cases, substantially reduced density to permit greater build without slumping. Thus a number of systems became available where the preblended powders were supplied together with a bottle of polymer dispersion, so that the site operative only needed to mix these two components to obtain a quality-controlled high performance repair mortar. Some manufacturers went a stage further and eliminated the separate bottle of polymer dispersion by incorporating a spray-dried polymer in the powder blend, so that only a measured quantity of gauging water needed to be added on site.

Not all formulators followed this route. One manufacturer eschewed the use of a polymer dispersion in the mortar and achieved the same objectives by full exploitation of cement chemistry. A shrinkage-compensated cement to give a fine measure of controlled expansion in the hardened state, coupled with the admixtures and specially graded fillers as above, achieved the desired result and proved acceptable to engineers, who at that time were still suspicious of polymer-modified materials. More recently this same shrinkage-compensated cement chemistry has been modified by polymer addition to give both high build and high strength alternative formulations with outstanding resistance to carbon dioxide and chloride ion diffusion.

Curing

Curing is a term likely to confuse the unwary: it has two completely unrelated meanings both of which are used in relation to materials for concrete repair. When the resin chemist talks of curing he is referring to the chemical reaction between the resin and hardener. The first transition stage in the reaction is when the liquid mixture becomes solid, or gels. Initially the gel may be quite weak, but as the reaction continues the material hardens further and achieves its ultimate strength, when it is said to be fully cured.

The cement chemist uses the term curing to denote the protective regime which must be provided around cement paste or concrete whilst it is hardening to prevent loss of water. This is necessary to ensure that the hydration reactions continue until much of the strength has developed and the pore structure has closed sufficiently to retain the essential moisture without external protection (Chapter 2, section 2.1.3).

The curing of concrete often receives scant attention. Adequate structural strength can often be achieved without careful curing, but the durability of the concrete may be seriously impaired. The core concrete may not be impaired because it is protected by the cover concrete, but the latter may be deficient in strength and, perhaps more seriously, have an open pore structure that gives ready access to carbon dioxide and water-borne chemicals, causing rapid carbonation.

If the near-surface concrete of large pours suffers as a result of inadequate curing protection, how much more likely is it that shallow concrete repairs, with no wet core concrete to feed them, will suffer dessication? Yet some contractors do not give it a thought. Cracking due to inadequate curing is one of the most common causes of failure in concrete repairs.

In their study of the characteristics of near-surface concrete, Dhir *et al* (1986) observed that microsilica concrete is more likely to suffer the effects of poor curing than OPC concrete is. Microsilica is a frequently employed constituent of formulated repair materials because of its beneficial effect in reducing permeability.

It is therefore essential that curing control, to a degree never practised in normal concrete construction, be exercised over cementitious mortar repairs. This is best done by spraying a curing membrane over the repairs as soon as the trowelling is finished. If a surface coating is to be applied over the complete repairs, the curing membrane must be selected to be compatible. Some polymer dispersions can be effective in this role and some resin-based curing membranes may be satisfactory, but the wax emulsions are obviously not suitable.

Fairing coats

If a concrete structure is to be protected by the application of an anti-carbonation coating, it is first necessary to prepare the surface. The usual requirements to remove all moss, algae, dirt, grease, laitence, old coatings etc. are fairly obvious. The need to remove any traces of mould oil and incompatible curing membranes should also be obvious, but is not always considered. Any suggestion that clean sound concrete might need re-surfacing before it can be coated is, for many engineers, an unacceptable elaboration. Yet even the best structural concrete will have blow-holes which are almost imperceptible until the surface is coated with a thin pigmented film. Then it will look like a painted colander and investigations at BRE have shown that these blow-holes let in carbon dioxide and cause carbonation to an appreciable depth. Whilst heavy-build coatings, often fibre filled, can bridge all the smaller blow-holes, the greater their capacity is in this respect the greater will be their surface texture and this may not always be aesthetically acceptable. With these coatings the larger blow-holes will require filling first and with thin-film coatings all blow-holes must be filled.

This is usually done by the application of a cementitious fairing, smoothing or levelling coat. Manufacturers vary in their approach to this matter; some offer a very fine textured material to be applied as thinly as possible by squeezing it into the blow-holes and scraping off all excess; others prefer a fine sand–cement-based material to be tight-trowelled to a thickness of 2–3 mm over the whole surface, including repaired areas. Some systems rely upon this layer for enhancement of carbon dioxide resistance.

These materials are of necessity polymer modified to give them good adhesion and thin layer integrity. Some require curing protection, whilst others contain moisture-retaining polymers to enable the cement fraction to hydrate without further protection.

Whilst the intention of the manufacturers is that these fairing or levelling treatments should be a preparation for the anti-carbonation coating, some contractors have been known to stop short at that final stage and simply use the fairing coat to improve the appearance of the concrete. The manufacturers will not usually give any assurances about the long-term durability of their fairing coats without the protection of the anti-carbonation treatment (Chapter 5).

4.2.6 Crack repairs

Cracks may develop in concrete from a variety of causes and before any attempt is made at repair the cause should be established and the necessity of repair should be seriously considered.

In the introduction to its report 'Non-structural cracks in concrete' the Concrete Society (1982) states:

It is fundamental that reinforced concrete cracks in the hardened state in the tensile zone when subjected to externally imposed structural loads. By means of appropriate design and detailing techniques these cracks can be limited to acceptable levels in terms of structural integrity and aesthetics. Concrete is also liable to crack in both the plastic and hardened states due to the stresses which it intrinsically sustains by the nature of its constituent materials. The construction industry does not readily accept that these intrinsic cracks are almost as inevitable as structural cracks, and concrete containing them may often be summarily and unjustifiably condemned.

Codes of practice have established acceptance limits for the width of cracks in structural concrete, but engineers frequently demand the repair of cracks well within these limits in the belief that they constitute faults and will impair the durability of their structures. This attitude is challenged in statements by the Concrete Society's Structural Engineering Steering Group that 'there is no firm evidence that cracks up to 0.5 mm wide lead to corrosion' and 'the design limit of 0.3 mm in our codes is therefore a visual limit, unconnected with corrosion in reinforcement'.

If, after consideration, it is decided that a crack must be treated, the repair philosophy should be established: is structural restitution required or is a cosmetic treatment sufficient? In the latter case a cement–polymer dispersion slurry worked into the face of the crack is probably the most satisfactory, but blending this in to achieve critical visual acceptance requires skill and practice.

There must be a realistic assessment of the possibility of achieving structural restoration of a cracked member. If the concrete is distressed again after repair it will crack again, often following the same route so closely that it is likely to be alleged that the repair has failed. If it is recognised that the crack is indeed 'live' then it will require proper consideration and treatment. Engineers will often call for the crack to be structurally repaired using a flexible resin. Although high strength flexible epoxy resin systems are available their extensibility is very limited (about 10%) and in moderately fine cracks they will behave in the same way as completely rigid resins. Even if the structural requirement is dropped, the use of an elastomeric system with high extensibilty is still a dubious proposition.

If one considers the painstaking efforts that are required to construct a properly designed movement joint in a concrete structure, it will soon become apparent that attempts to convert an accidental crack into an effective movement joint have small chances of success. Cracks in concrete are so irregular that if even a comparatively straight one is closely examined it will be seen that, as the crack opens, the strain to which the infilling elastomer will be subjected will vary tremendously and be extreme in places. It is usually best to repair the crack fully and then to cut a deep chase as a crack inducer and seal it at the face in the normal manner. If the width of this seal is calculated to be within the performance of the sealant for the anticipated movement a fully effective repair will be achieved and any further movement will be properly accommodated.

Epoxy resin injection

Although the injection of cement grout into a crack is often the first option an engineer will want to consider, it has very limited chances of achieving a satisfactory structural repair. Investigating the injection of cracks up to 5 mm wide in reinforced concrete beams, Dubois (1981) found that cement grout could not penetrate or was totally ineffective where the crack narrowed below 0.75 mm.

In both short- and long-term loading tests he found that cracked beams repaired by epoxy resin injection deflected less than those injected with cement grout, the greater penetrability of the resin more than compensating for its lower elastic modulus. No doubt the higher shrinkage of the cement grout contributed to its poorer performance.

Investigations by Chung (1975) and Hewlett and Morgan (1982) showed that epoxy resin injection can fully restore the strength and stiffness of cracked reinforced concrete beams. In some cases the repaired beam has been found to be slightly stiffer than before. This is because the resin is able not only to rebond the concrete but to bond the concrete more firmly to the steel than the original cementitious bond. It has been shown by the author that the restored load-bearing capacity of resin-injected columns was maintained after half an hour of fire testing.

It is generally not practical to inject resins into cracks less than 0.2 mm wide at the face, but provided the crack is of sufficient width to make resin entry feasible, the low viscosity resin systems are then able to penetrate into the finer depths of the crack down to limits measured in microns. Even when such resins are made thixotropic to prevent free drainage from incompletely sealed cracks, they can reach the extremities of cracks well below 50 μm.

At the other extreme, unfilled epoxy resins are not usually recommended for injection into cracks greater than 10 mm in width because of the risk that they develop too much exothermic heat during cure and subsequently con-tract. Rather than thickening the resin with a heavy loading of fillers and so reducing penetration into finer parts of the crack, it is better to pack the wider parts pneumatically with single-sized fine aggregate and then to inject with unfilled resin.

Cement grouts

Cement grouts can of course be used for the repair of wide cracks but, to avoid the penalties of high shrinkage, proprietary grouts based upon shrinkage-compensated cement systems should be considered. Whilst many cementitious grouts contain expansive agents, a clear distinction should be made between those which expand in the fluid state and those which produce expansion after setting. The former are useful in developing a little pressure by the genera-tion of fine gas bubbles (hydrogen from aluminium powder, air from activated carbon or nitrogen from metastable organic compounds), which is a useful aid to penetration, but only the latter, which expand in the hardened state by a limited and carefully controlled ettringite formation, can truly be called shrinkage compensated. They can achieve a fully monolithic union with the fractured host concrete. Those containing fine aggregate and sands are suitable for filling cracks from 10 to 100 mm in width, and by incorporating 6–10 mm aggregate they can be used structurally to fill much larger voids but then are more likely to be used for reconstruction than crack filling. This leads to discus-sion of another group of materials, the superfluid microconcretes.

4.2.7 Superfluid microconcretes

Excellent though many of the formulated repair mortars are, in practical terms there are certain limitations to their use. They are ideally suited to the hundreds of patch repairs that may be required on a typical building or structure suffering deterioration as a result of variable depth of cover and irregular compaction. In some situations, however, a more wholesale replacement of the facing concrete is required and a more efficient method than hand placing becomes an economic and logistic necessity. The congestion of reinforcement can also be another factor precluding effective repair by hand placement.

These situations call for a concrete-pouring technique such as was appropriate to the original construction, but the narrow section (typically 50–100 mm) extending over a pour perhaps 3 or 4 m in height would make it impossible to compact a normal concrete. The answer lies in the use of fluid cement grouts bulked out with small aggregate.

To base such a mix on a simple cement grout would be to invite severe shrinkage. Superplasticizers must be used to achieve fluidity whilst keeping the W/C down to 0.4 or less. As was discussed in section 4.2.6, two-stage shrinkage compensation can be built into the formulation: gas expansion in the fluid stage, and further slight expansion after the repair has hardened can prevent distress over the ensuing weeks as full drying occurs — hence the use of the expression 'pre-shrunk' concrete at the beginning of this chapter.

The quantity of aggregate which can be incorporated may be limited by the method of placing. If a simple pouring technique is used, 10 mm aggregate may form 50% of the total dry materials, but if pumping is preferred the aggregate size and quantity will need to be reduced to suit the limitations of the pump.

The term superfluid microconcrete has been coined for these materials because they are so far removed from normal or even collapsed-slump concretes. Their consistency is more like that of lentil soup and engineers have been known to walk away in contempt, firmly convinced that such a watery-looking mix could never reach 60 N/mm^2 compressive strength. The test cubes have invariably proved them wrong and when the shutters have been removed they have been greatly impressed by the superb finish obtained.

By careful control of the density and rheology of the mix, the aggregate should remain uniformly suspended throughout the depth of the pour.

When this technique is used for the repair of bridge structures under the control of the Department of Transport, the contractor is required in any case to demonstrate the performance of the material and the proposed mixing and placing equipment in a large-scale laboratory trial. For this a Perspex-topped horizontal soffit mould 750 mm square and 50 mm deep with appropriately spaced reinforcement is filled from one side with only a 75 mm head of material. This is detailed in the Department of Transport specification BD 27/86 (Department of Transport, 1986) which also details the flow test to be performed upon each batch of material.

Superfluid microconcretes do not need to be associated with any other products to present a 'repair package'. There is less justification for anti-corrosive treatment of the steel reinforcement, since this type of concrete (as distinct from a repair mortar) fully wets and envelops the steel and the large areas involved lessen the likelihood of only part of a bar being encased in the fresh concrete. There is no need for any bond coat because the new material is fully supported in place by the shutter for several days, during which time a strong cementitious bond is able to develop. Site experience, as well as laboratory tests, bear testimony to this.

Although the shutter gives protection during the earlier stages of curing it is still important to apply a curing membrane to control moisture loss immediately after striking the shutters.

4.3 Repair techniques

The practical aspects of using the materials studied in the preceding sections are now considered. The preparation of the concrete is reviewed first and then the various stages in carrying out the repair using hand-placed mortar systems, the recasting process using superfluid microconcretes, crack injection and, finally, sprayed concrete.

4.3.1 Removal of defective concrete

If a full and thorough diagnostic survey has been carried out as discussed in Chapter 3, the engineer will be able to define fairly precisely the extent to which concrete is to be removed, but some contracts are undertaken without extensive analysis and the decisions may be taken on the spot by the engineer as the deteriorated reinforcement is exposed. Carbonation tests are sometimes carried out *in situ* as the work progresses, but it should be appreciated that carbonated concrete is not in itself necessarily defective; it is only when the carbonation front reaches the steel that the question of removal arises. As long as the steel can be shown to be encased in even a few millimetres of good alkaline concrete, carbonation of the rest of the cover depth does not in itself put the steel at risk, provided that steps are immediately taken to prevent any further advance of the carbonation front.

Whenever corrosion of the steel has occurred the concrete should be removed around the entire circumference of the bar and this removal process should be continued for at least 50 mm along the bar beyond the furthest extent of corrosion. This work is normally carried out using hand-held pneumatic chisels, but occasionally very hard concrete is encountered which requires the use of heavier equipment. Very high pressure water jetting may be useful on these occasions, particularly when dealing with slender members where the use of heavier tools could cause some structural concern. When high pressure water jetting is used it will usually be necessary to supplement this with further work by pneumatic chisels to remove concrete behind the reinforcement bars which cannot be reached by the water jet, and also to give a sharply defined perimeter to the repair.

This detailing of the periphery of any patch repair is very important. It should be achieved by disc cutting to a depth of 5–25 mm (determined by the minimum depth to which the repair composition can be applied) unless the repair composition has been specifically formulated to permit its application down to a feather edge. Some engineers prefer to mark out the line of this disc cut before breaking out commences, so that they can control and measure the amount of concrete removed, but in other cases the extent of the removal is decided on the spot as exposure of the rusting steel progresses and the disc cutting is then carried out after the bulk of the concrete removal. In either case finer tools should be used to break out the concrete close to the disc cut to avoid any damage to this edge.

The disc-cut face will itself require some further preparation, either by wire brushing to loosen embedded dust and roughen the face or, in conjunction with the treatment recommended in the next section for cleaning the steel reinforcement, grit blasting. In this case it will be necessary to hold a board against the outer face of the concrete to protect the arris from erosion.

If it has been necessary to use heavy tools for the breaking out, a careful examination should be made of the exposed concrete surface and, if fracture of embedded aggregates has occurred, these should be subsquently removed with lighter tools.

Finally, it is necessary to remove all dust from the prepared concrete, although in most cases this will have to be done after preparation of the steel reinforcement which has been exposed. The use of a good industrial vacuum cleaner with small brush heads is recommended, but the contractor's preference is usually to distribute the dust widely by blasting with compressed air. If this method is used the hoses should be kept clean and the filters for removing oil and water from the air should be regularly serviced.

4.3.2 Cleaning the steel reinforcement

If steel reinforcement has been exposed because it is rusting, it is essential that all these corrosion products be removed before repair work progresses.

The need for rust removal is often questioned when there are no chlorides involved, the argument being that rusty steel is used in new construction. There is a big difference. There should only be a thin patina of rust on new reinforcement; if it is any heavier the engineer will usually require it to be cleaned off. This very thin layer of rust is converted to the passivating oxide film by reaction with the cement paste when it is encased in concrete. If there is a heavy rust deposit, as must be the case if the passivating film has been destroyed and corrosion has necessitated repair, the cement paste of the replacement concrete or mortar will not be able to convert it all and a passivating film will not be developed on the steel. Grit blasting is undoubtedly the most effective way of cleaning the steel, but it can cause severe problems with dust and noise. The dust can be eliminated by the injection of a small quantity of water into the air stream; the ensuing sludge is removed with a jet of water.

It is sometimes assumed that when high pressure water jetting is used for cutting out the concrete this will automatically clean the corroded steel. It may be necessary to introduce sand into the water jet to provide sufficient abrasion to clean the steel properly.

The reluctance of all parties to use these expensive and messy preparation methods is understandable, and some contractors will try to get away with the use of wire brushes. This is completely futile as it only polishes the rust, often to a patina of deceptively metallic appearance. A quick rub with a carborundum stone will soon break through this and expose the powdery rust beneath. Where rusting has been caused by the presence of chlorides, deep pitting is likely to occur and it will be necessary to use a powerful jet of water to remove soluble corrosion products from these pits, even after grit blasting.

4.3.3 Protecting the steel reinforcement

Freshly cleaned steel will begin to rust within a few hours if it is not protected. If it is simply to be encased in concrete there is no problem because, as noted in the previous section, that rust will be converted by reaction with the alkaline cement paste.

However, if the steel is to receive any of the other protective treatments discussed in section 4.2.5 there should be the minimum of delay, because these anti-corrosion treatments are generally most effective only on clean bright steel. In most circumstances four hours is considered the maximum time which should elapse between grit blasting and the application of a protective coating to the steel. In a maritime situation, however, where chlorides are carried in the atmosphere, it is better to keep the exposure of the clean steel to an absolute minimum, say 20–30 minutes.

The various alternative anti-corrosion coatings were reviewed in section 4.2.5. The majority are two-part materials and need to be mixed immediately before use. Some of the polymer–cement slurry coatings are simply proportioned with a measuring cup provided. Some of the epoxy systems are similarly volume batched on site, but the more foolproof systems are supplied pre-weighed, requiring the entire contents of the resin base and hardener containers to be thoroughly blended. Composite containers in which a separating membrane is ruptured to release the hardener into the resin are a further sophistication. Whatever the method of batching, thorough blending, with careful scraping of the sides and bottom of the container, is most important.

Applying the coatings is a straightforward painting operation, although congestion of reinforcement can complicate matters. Small cranked brushes are useful for reaching behind the bars. A single full coat of the polymer–cement slurries and anti-corrosive epoxy paints is adequate, but the heavy solvent-free epoxy coatings usually require two coats. Although fresh concrete or cementitious mortar can bond well to some wet epoxy coatings, it would not be appropriate here because the film would be too severely disrupted when the mortar was applied and so it must be allowed to harden without disturbance. Cement paste does not bond well to the cured resin, however, and so a second resin coat is applied and dashed with fine sand to provide an excellent key for the repair material.

Probably the simplest protection system to apply is a one-pack zinc-rich epoxy coating. Because the epoxy groups have already reacted in the preparation of the binder resin, this is simply a solvented system brushed on like paint and left to dry for an hour.

4.3.4 Using a bond coat

The reason why a bond coat, primer or adhesion promoter may be needed was discussed in section 4.2.5. It is now necessary to consider the methods of mixing them and applying them to the prepared concrete substrate.

As already stated (section 4.2.4), polyester mortars do not usually need any bond coat or primer, but epoxy mortars almost invariably do. This is normally a two-component system (base and hardener), usually containing a solvent to facilitate application as a very thin film and to improve wetting-out of the concrete with some degree of penetration. Thorough mixing of the two components is easily achieved because of their low viscosity, and the application of the mixed primer by brush presents no difficulty.

The bonding requirements of cementitious materials are rather different and varied. With the superfluid microconcretes a bond coat is not usually required (section 4.2.7). It is only necessary to saturate the substrate thoroughly, preferably by sealing the shutter in place and filling it with water.

A bonding coat is of course essential when using trowel-applied cementitious repair mortar. As already discussed (section 4.2.5) these are usually polymer based. The substrate should be pre-dampened, preferably thoroughly saturated, and excess water removed. Many of the polymer–cement systems require the polymer dispersion to be diluted and some are supplied in dilute form. It then remains for the powder to be smoothly blended with the dispersion in the proportions specified by the formulator or by using complete pre-weighed packs as supplied.

For separately purchased ingredients a typical recommendation would be 1 volume of water to 1 volume of polymer dispersion and 3 volumes of OPC.

In the case of the one-pack polymer bond coats where no cement is added, there is no mixing to be done and the material is simply added straight from the container. Whichever type is used, the technique is the same: to work the material firmly into the surface of the dampened substrate, taking care not to progress too far ahead of the mortar application, so that the mortar is always applied to the wet bond coat. Should some mishap prevent this and the bond coat is allowed to dry, it should be wet scrubbed off and the procedure restarted. This of course does not apply to those polymer-only systems which can be reactivated after drying, when a second coat may be applied.

It is a different story when the heavier bodied epoxy bond coats are used in conjunction with cementitious materials. These are only used in special circumstances, such as the repair of concrete affected by alkali–silica reaction (ASR) where, after removal of all deteriorated concrete, it is imperative that no water be introduced when carrying out cementitious repairs. There are other occasions when soaking of the concrete substrate is undesirable, or when a barrier is required to prevent the possible movement of moisture and/or salts between the substrate and repairs, or simply to incorporate a damp-proof membrane with structural capacity.

Epoxy bonding agents are preferred for the latter situations where the integrity of the coating as a barrier is crucial. Two coats should be applied, and the first, functioning as the membrane, should be allowed to set but not fully cure before the second is applied. The coat into which the fresh concrete or mortar is compacted will be breached by aggregate particles and so cannot function alone as a barrier membrane. Because they are more viscous filled materials, these two-part epoxy systems need more effort and diligence in their mixing and a Jiffy mixer (Fig. 4.3) which scrapes the sides and bottom of the container is preferred to stirring with a rod or stick. Some manufacturers colour the base and hardener in contrasting colours so that streakiness due to inadequate blending is easily seen.

Whilst a strong brush may often be the easiest way to transfer the mixed material from the can to a vertical surface (on concrete decks it may simply be poured as a ribbon to distribute it roughly over the appropriate area), it is vital to ensure that it is then firmly worked into the roughened concrete surface by a scrubbing action with a stiff-bristled brush. The fresh concrete or repair mortar must then be applied while the epoxy coating is still wet and sticky. This generally presents no problem as the well-formulated epoxy bond coats have an open time of many hours. If by any misfortune there is a hold-up in the cementitious application the tacky bond coat may be heavily dashed with clean dry sharp sand and left to harden, providing a key for a repeat operation at a later date. Failure to do this will result in a hard shiny epoxy coat which will require complete removal by grit blasting before operations can be recommenced.

Fig. 4.3 Jiffy mixer.

4.3.5 Mixing mortars and concretes

Epoxy materials

The blending of epoxy resin base and hardener has already been considered in the previous section and this is again the first stage in the preparation of an epoxy mortar. There is some temptation to do the blending in the large mixer required for preparing the total mortar but it is not to be recommended; mixers of all types are only efficient at their optimum loading, and since the resin binder accounts for only one-third to one-fifth of the final bulk it is unlikely to be thoroughly blended running around the bottom of a large mixer. In any case, most of the mixers used for this work rely on the shearing of the paddles through a heavy mass of aggregate for their effect. So, the binder should be prepared first in a smaller vessel. Most formulators supply pre-weighed packs with sufficient spare capacity in the base resin can for the hardener to be added and properly blended.

The contents should then be transferred to the larger mixer together with the aggregate or filler. Some manufacturers advocate pre-mixing the filler and then adding the binder, but if the condition of the mixer is less than perfect there is a risk of finding a layer of dry compacted filler on the bottom at the end of the mixing period. Putting the binder in first does not give the reverse effect as the filler will usually sink into the binder and a homogeneous mix is more easily achieved.

The equipment required for this is a forced-action mixer in which the drum of material is rotated against adjustable static or rotary tools. It is important to keep the drums and tools clean and in proper adjustment, so that the bottom and sides of the vessel are thoroughly scraped. If the drum is kept clean there is no need to hammer it to remove the contents, but if for any reason the drum becomes distorted it should be replaced. It is always a wise precaution to keep a spare drum and, on contracts where there is continuity of demand, the use of two drums and a handling trolley makes for more efficient employment of the mixer.

Small batches of up to 3 litres total can be mixed by hand if great care is taken, although BS 6319 (1983) only permits hand mixing for up to 1 litre total for test purposes. A pair of stout polythene buckets is ideal for this and a good mixing spatula can be made from a bricklayer's trowel cut off squarely at the end with the sides trimmed parallel to make a 2 inch wide blade. When all the batch appears to be thoroughly blended it should be tipped into the second clean bucket, the first one should be scraped out and a further period of mixing given. This may sound a little tedious but it is quicker to do it than describe it and the overall effort is probably less than that required to achieve a really uniform blend in a single bucket.

Cementitious materials

A good forced-action mixer is to be preferred for the preparation of cementitious repair mortars and superfluid microconcretes. Free-fall concrete mixers are generally not recommended because they cannot impart sufficient shear to overcome the initial water repellency of many of the formulated powders or to blend in the lightweight fillers that many of them contain. Rotating pan mixers with adjustable static and rotary tools are needed to give high shear to the materials and more or less continually scrape the sides and bottom of the pan to ensure homogeneity. Although the motor driving the pan does most of the work, some effort is required of the operator in working the plough or stationary scraping blade back and forth between the sides and centre of the rotating pan.

For small-scale works where single-bag (15–25 litre) mixes are required a heavy duty power drill fitted with a properly designed helical mixing paddle will suffice. When mixing mortars, considerable effort is required from the operator to work the rotating paddle up and down and progressively around the mixing vessel to achieve a sufficient degree of mixing. This can only be satisfactory where batches are required on an intermittent basis, for otherwise operator fatigue is likely to result in skimped mixing. Superfluid microconcretes are a different proposition and are quite easy to mix by this method. The vessel should be a sturdy drum of such dimensions that the mix will fill it to a height of 1–1.5 diameters.

Because most formulated materials contain very efficient plasticizers they are highly sensitive to water content. It is therefore essential that the gauging water be carefully measured. In some cases the site operative has no responsibility in the matter: the gauging liquid, a diluted polymer dispersion, is supplied in a separate container for each bag of powder. Some formulators supply the polymer undiluted and the additional water is measured on site, usually by refilling the same container. This may lead to error, particularly if the containers have become distorted during storage. When the bag of powder contains all the ingredients including a spray-dried powder, the total quantity of gauging liquid (water) must be measured on site. This is very simple provided that a little common sense is exercised. Measuring vessels should be as tall and narrow as possible and should be stood on a firm surface whilst being filled. A slot or hole in the side of the vessel is more foolproof and sensitive than a painted line, its precise position being determined by measuring the desired quantity of water into the vessel, preferably by weighing (the graduations on plastic buckets are quite unreliable). The number and capacity of vessels should be sufficient that the total water requirements for a mix can be measured and checked before any of it is transferred to the mixing vessel. It is quite incredible how some contractors will jeopardize the consistent quality of their material by meanness in this respect.

The gauging liquid should be poured into the mixing vessel and then the powder should be added with continuous mixing. In some cases, particularly with the superfluid microconcretes, it may be advisable to keep back a proportion of the water so that the initial mix is on the stiff side, to generate sufficient shear to give thorough mixing, and to then add the remainder when all the powder has been wetted and uniformly dispersed. Particular attention should be paid to the formulator's instructions regarding the time of mixing. The solvation of plasticizers and the redispersion of powder polymers may not be instantaneous. This may cause the mix to appear initially too dry although uniformly mixed. The temptation to add more than the specified amount of water must be resisted firmly and the mixing continued. After another one to two minutes the consistency may often change dramatically to give the desired workability (excessive mixing after this may cause air entrainment and loss of strength). In some cases the formulator may recommend allowing the prepared mix to stand for ten minutes to allow further development before being used or even re-gauging with a few drops of water after that time to achieve the right consistency.

The 'right' consistency for a mortar will to some extent depend upon the method of application (whether by gloved hand or trowel) and should, in any case, be in accordance with the formulator's guidelines. The tendency is to make the mixture on the dry side in the belief that this will give the best 'build' characteristics, but this does not necessarily apply. A slightly wetter mix will often have more cohesion and, if the formulator has got the rheology right, can be less likely to slump and 'belly' than a dry one. The dry mix is also more likely to dry out too quickly and give poor strength owing to inadequate hydration. The author's preference is for a consistency such that a ball of mortar 'fattens' or glazes on the surface when tossed back and forth in the hand or from trowel to hawk three times.

The 'right consistency' for a superfluid microconcrete is best measured in a flow trough, such as that specified by the Department of Transport (1986) although that document does not adequately detail the method of use. For consistent results the trough should be thoroughly washed and completely inverted to drain for one minute; the edges should be wiped with a damp cloth before the trough is reverted and placed on a level surface and checked with a spirit level. Method statements for specified contracts are based on this document and give various minimum flow requirements, but there is also a need for an upper limit to flow or a separate check for bleeding and settlement of the aggregate. When the appropriate flow characteristics have been established arrangements can be made to dispense the required amount of water consistently, although there are some grounds for allowing a little latitude for final adjustment by 'experienced eye' to compensate for slight variations in water demand as cement and natural sand sources fluctuate.

4.3.6 Placing

Mortars

Although resin mortars and cementitious mortars are very different, both chemically and in their 'feel' or response under a trowel, much of the detail of application is the same, so to avoid unnecessary repetition they will be dealt with under one heading.

After priming the substrate a period of drying may be necessary if, as is most common for resin mortars, the primer is solvent based. It is in any case better to wait until the primer develops to a near-dry tacky condition, and the mortar can be prepared during that interval. Polymer primers for cementitious mortars similarly need a drying period, but polymer–cement priming systems or 'bond coats' develop immediate tack and no delay should be permitted before placing the mortar.

The mixed mortar should then be packed into place a little at a time, taking particular care to compact the first layer firmly into the primed surface. The best method of doing this will depend on the situation; it may be simple to pack the material into the cavity if it is bounded on all sides by concrete in one plane, but when reforming an arris it will often be necessary to use some form of support to one face. If the repair is shallow this may just be a hand-held board or a trowel, but more substantial repairs will require a fixed shutter. Similarly, when repairing a soffit it may be necessary to use a shutter if the depth is more than can be applied overhead in one pass.

Hand-held supports may be slid away when no longer required, but fixed shutters will require some form of release treatment. Whilst wrapping in clean polythene or self-adhesive polythene tape are effective methods, for the best finish a steel plate or a lacquered or laminated board should be used. For use with a resin mortar this should be well treated with a hard grade wax polish and then lightly sprayed with a silicone mould release agent. Shutters for cementitious systems should be treated with a light coat of mould oil or chemical mould release agent.

When filling an open cavity, a handful of mortar can be consolidated into a ball using the gloved hand and then firmly pressed out at the back of the cavity. Successive handfulls should be similarly placed, taking great care to squeeze out air between layers and to pack the material well behind reinforcing bars. In vertical and overhead situations there will be a definite limit to the thickness that can be applied without slumping. This will vary considerably from one formulation to another according to the density, wet cohesive strength and rheology of the mix. Each formulation will represent a compromise between these 'build' characteristics, cost and the physical properties of the hardened product. If the material does slump it should be removed and replaced to a lesser thickness. Attempts to push or work it back into place are likely to result in hollowness because of the impaired bond.

If these limitations on build mean that the repair cannot be completed in one operation, the first layer should be finished to a reasonably uniform face and lightly scratched to provide a key for the following layer. Scoring with the point of a trowel is not recommended: the depth of scoring is uncontrollable, can induce cracking and does not roughen an appreciable proportion of the surface area. Light scratching with fine wire tines about 5 mm apart (like a miniature grass rake) in a 'wiggly' pattern is far more satisfactory. When this layer has hardened the scratched surface should be reprimed and a further layer of mortar applied.

In some situations, such as floor repairs, it may be appropriate to use a trowel to apply and compact the mortar. If so, it should be a wooden one and fairly small to give maximum compaction.

Whether layered, built up in one pass or backed behind a shutter, the open face should be consolidated slightly proud of the surrounding concrete face and then struck or cut off flush. The surface should then be 'closed' using a steel float. With resin mortars this will need to be kept clean by wiping with a rag moistened by a little solvent. The trowel or float should never be wet with solvent or the durability of the finished face may be impaired. Cementitious mortars are more user friendly and do not cling to the tools, so no special precautions are necessary.

There is one very big difference between resin-based and cementitious mortars – the need for careful protection of the newly applied mortar. Resin mortars need none, apart from avoiding mechanical damage. Cementitious mortars are very sensitive to moisture loss (section 4.2.5) and in strong drying conditions the mixed mortar may need protection by covering the container with a wet rag when being hoisted to the scaffold, or during use if it is not all to be used in a few minutes. The work must be shaded from direct sunlight until fully cured and the provision of wind breaks may be necessary. The urgency of providing curing protection to the finished (or part-finished) work cannot be over-emphasized.

When it becomes necessary to build up a cementitious repair in more than one application, the first and any intermediate layers must receive careful curing protection (after scratching to provide a key). This may mean draping wet hessian over the face of the repair and then protecting this with securely fixed polythene, or the spray application of a curing membrane. The latter method can only be used if the curing membrane is offered as part of a formulated package to be compatible with the bonding agent required for the next application. The delay between successive applications should be no more than is required for hardening of the underlayer, or at the most overnight. In this way each layer provides the ultimate in curing protection for the preceding one and maximizes the inter-layer bond strength.

The final layer is the most vulnerable to curing problems. Where the area of a patch exceeds $0.75\,\mathrm{m}^2$ (or a 1 m run of narrow repair) it is advisable to finish the work $0.5\,\mathrm{m}^2$ at a time and apply curing protection before continuing. Waiting to finish the whole patch before protecting with a sprayed curing membrane or wet hessian and polythene may be too late, although the consquent shrinkage cracks may not manifest themselves for several weeks.

Reference has already been made in section 4.2.5 to the need to consider the effect of the curing method upon any subsequent finishing technique. Some formulators offer curing membrane spray compounds which are compatible with their surface coating systems. Of course, if there is to be no finishing operation any conventional curing membrane may be used. It is nevertheless advisable to check the coverage rate necessary to achieve the published curing efficiency for any particular product. Since these materials are normally formulated for and tested in horizontal application, it may be found that the recommended quantity cannot be applied in one pass in the vertical. It may be necessary to allow some minimal drying time before applying a second coat.

Even a high efficiency curing membrane may not be sufficient in strong drying conditions on an exposed site. It may still be advisable to supplement it by taping polythene over the work as soon as it is complete.

The need to conserve moisture in the repair mortar also means that any fixed shutters should be of metal or plastic or, if timber, receive two coats of polyurethane shutter paint. They should also receive a very thin coat of mould release.

Superfluid microconcretes

These materials can be placed by pumping or pouring and the real secret of success is to place the material as fast as possible. This means setting up to mix and place on a continuous basis, yet some contractors attempt fairly large works with a single drill and paddle producing batches at five to ten minute intervals. At the other extreme, large capacity mixers with integral holding tank and heavy duty pump can place large quantities in a very short time, so these require good organization to ensure continuity of work once they have been set in motion. For many jobs such sophistication is not justified and good results can be achieved by a team with two drill-and-paddle sets or a Cretangle mixer with a spare drum. It is always wise to have a standby mixer at the ready in case of breakdown in the middle of a large pour.

When the repair section exists in only one plane, i.e. a simple vertical section, it is sufficient to pour the material in from the top of the shutter; it will not suffer the severe segregation which occurs with normal concretes and causes the all-too-familiar 'boney' kicker joint responsible for so many water seepage and corrosion problems. It is better to allow the material to fill the shutter by continual addition at the top than to keep feeding from the bottom and trying to displace the first-poured concrete all the way to the top. Although these materials are formulated to retain their flow properties over some 30 minutes, this is only maintained with regular agitation. When the mix 'stagnates' in the mould it develops 'false body', making it difficult to displace it upwards if feeding from the bottom. This movement of stiffer material and the bursting through of the fresher material tends to give more flow patterns and blow-holes at the shutter face.

If the repair is more than 1.5 m wide it is advisable to have two pouring points being fed simultaneously. It is also necessary to consider venting the top of the cavity if it is enclosed. It is advisable to slope the top edge towards the front and laterally to bring it to a slight rise at the extremities and between pouring points. Chutes should be formed at the top of the shutter at the pouring points and they should be extended to a height of about 150 mm above the vent points (Fig. 4.4). The latter can be small chutes 50 mm wide extending 10 mm above the top of the cavity. As soon as the new concrete has set these chutes should be removed and the protruding casting dressed back. Of course,

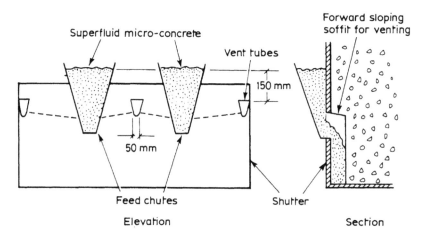

Superfluid micro-concrete

Vent tubes

Forward sloping soffit for venting

150 mm

50 mm

Feed chutes

Shutter

Elevation

Section

Fig. 4.4 Placing of superfluid micro-concrete in a vertical section.

if the repair extends to the top of the member the shutter may be open topped and none of these elaborations will be necessary.

Soffit repairs If the repair is to a soffit the supply arrangements will be quite different. If it is the soffit is to a moderately thin slab it may be possible to drill feed and vent holes right through the slab. If the area exceeds 1.5 m² it would be wise to have a second feed hole. The material should not be poured directly down the holes but through a loose fitting PVC pipe, a rainwater pipe say 50 mm or larger (with a hopper on top), extending 0.5 m above the top of the void and passing through a hole in the slab 20 mm larger in diameter to provide venting at the same point (Fig. 4.5). Because of the difficulty in making much of a slope over an area of cut-out soffit, it is necessary to make a number of small vent holes, preferably by drilling upward through the slab from observed higher spots.

If drilling through the slab is not acceptable or not feasible, the soffit repair must be fed by pipework sealed into the face of the soffit shutter. This can then be fed by gravity if the pipework can be led to a position where a feeder hopper can be fitted 500 mm above the highest point of the cavity; otherwise it must be fed by pump.

If the soffit repair extends to the edge of the member and is contiguous with a face repair, the complete repair may be poured as one, but the soffit should be plumbed and filled as just described and then the face section poured from the top.

Shutter design In all these situations the shutter must be stout enough to resist the full hydrostatic head of the concrete without deflection and must be securely clamped against the face of the host concrete so as to withstand this pressure

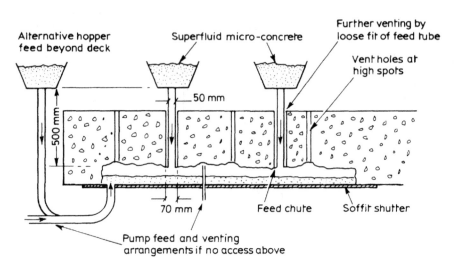

Fig. 4.5 Placing of superfluid micro-concrete to a soffit.

without grout loss. A bead of mastic or a strip of compressible foam can be applied around the face of the cavity before clamping the shutter to improve the seal.

The shutter face coming into contact with the concrete should be foil laminated or receive two coats of polyurethane shutter paint. The concern here is not so much the texture of the cast face as the need to minimize the passage of moisture through the shutter. The face should also receive a light coat of chemical mould release.

The host concrete must be saturated before pouring the superfluid microconcrete, again to prevent loss of water into the substrate. This is best done by filling the shutter with water, checking the joints and seals for leaks and keeping it full for a couple of hours or more. Sometimes the contractor may object that the water supply is limited and he does not want to waste so large a volume.

The alternative is to set up a mist spray and keep the entire face saturated in that way for a similar time. Before the pour commences the water must be completely drained away. This requires the provision of one or more drainage points drilled (from the inside face before fixing the shutter) at the lowest points. Puddles of water left in the shutter may be displaced if the grout flow is steady from one side, but if there is turbulence or the mix is poured in from the top the excess water is likely to dilute the mix and cause bleeding and/or segregation. When all the material has been placed any exposed material, e.g. at the top edge of a vertical pour, should be protected with polythene sheeting.

Because this form of repair has something in common with traditional construction and even looks like a normal concrete pour, there is a tendency for the unwary contractor to give his normal attention to curing, which means leaving it to take care of itself! As with trowelled repairs, because there is no substantial body of wet concrete behind the 50–100 mm depth of repair, the new concrete needs very prompt and effective protection. The shutters should be stripped as early as possible, consistent with the material developing adequate strength, and a high efficiency curing membrane should be sprayed on.

4.3.7 Crack injection

Although epoxy resin systems can be shown to have good penetration into cracks tapering down to less than 50 μm in width, it is not usually practicable to inject fine cracks unless the resin can be introduced at a point where the crack is 0.2 mm wide or greater. Polymer dispersion (SBR or acrylic 'emulsions') are more efficacious where the crack is less than 0.2 mm at its widest.

Crack preparation

Much ill-conceived advice has been written about chasing out cracks and drilling holes to fix injection tubes. Although the proponents of such techniques might get away with the semblance of a repair if the crack is of ample width, coring, demanded by the dubious engineer, will probably reveal that the resin has not penetrated the finer parts of the crack because debris from the drilling or chasing has blocked the crack. If the crack is open at the surface it is best to fit flanged injection tubes (with adhesive-coated backing washers) over the crack without any mechanical preparation. A wipe with solvent to remove surface dust is all that is required.

If the crack is blocked at the surface then some further preparation is necessary. In the author's experience the most practicable expedient is vacuum-flushed drilling. The equipment for this is now commercially available and is simply a rotary manifold which accepts hollow-stemmed tungsten-carbide-tipped drill bits and fits into any standard percussion drill, or better still a rotary hammer. A small vacuum cleaner connected to the side outlet of the glanded manifold draws air up the drill tube from holes immediately behind the cutting tip. The debris is thus drawn immediately from the cutting face leaving the crack perfectly clean. By contrast, standard fluted non-flushed drills pack the debris tightly into the crack.

Because one can seldom be certain that the crack travels into the concrete at right angles to the surface, it is prudent to drill the holes at an angle of about 45° from a point about 25–30 mm away, to intercept the crack about 30 mm down from the choked surface. Plain copper tubes (60 mm lengths of

8 mm soft copper are ideal) can then be sealed into the 10 mm holes with a filled polyester paste. The next step is to check that the drilling has been successful and that the tube now connects with the crack. If one cannot blow air (by mouth) through the tube there will be no hope of pumping resin through. A piece of laboratory rubber tubing will suffice, but it is worthwhile fitting a bubbler (Fig. 4.6) into the line to give visual indication of the passage of air if the crack is very fine. If a good air flow is not achieved, another hole should be drilled nearby, this time from the opposite side of the crack, and another tube fitted.

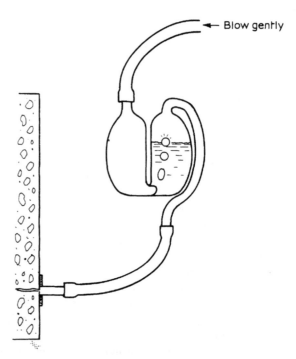

Fig. 4.6 Bubbler, made from wine-making air-lock. Stream of bubbles through the water indicates a clear passage through injection point and through crack to an outlet.

Whichever method is used to fit the injection tubes, they should be bonded into place with a quick-setting filled polyester paste. This should be formulated with a soft filler so that it is easy to rub down with a carborundum stone. The crack between the injection points should also be sealed with this paste. It is not necessary to build it up into a poultice as some manufacturers suggest; rather, it should be knifed into the crack and scraped off flush with the surface. The day after injection, surface mounted tubes can be knocked off, drilled tubes cut off cleanly with a sharp cold chisel and the face of the crack dressed down with a carborundum stone. Unless the cracking has caused a step in the face of the concrete, the repair crack should be almost invisible.

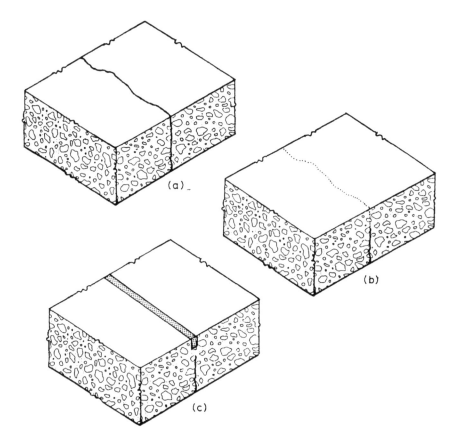

Fig. 4.7 Repair of a 'live' crack. (a) Cracked slab. (b) Crack partially rebonded by resin injection to controlled depth (see Fig. 4.8). (c) Crack-control slot cut and filled with flexible sealant. Any further movement will propagate crack to base of slot, thus permitting movement but keeping crack sealed against ingress.

Placing of injection points

The spacing of the injection points depends upon a number of factors. The first is the depth to which the resin is required to travel; in many cases it is the full depth of the crack, but there are circumstances where a limited depth of rebond is required, e.g. in the repair of a live crack (Fig. 4.7). If the crack extends the full depth of a member but the far side is inaccessible, it is necessary to limit the travel of the resin to approximately the depth of the member, otherwise a great quantity of resin may be wasted and the crack may never be filled. The technique for achieving this is shown in Fig. 4.8 and depends upon the use of a thixotropic resin. The spacing between the injection points should be five-sixths of the depth that the resin is required to reach. It can be seen from the sketch that a free-flowing resin could go to waste (and coring of floor slabs treated by gravity feeding of such resins has proved the point), whereas a thixotropic resin will not drain under the influence of gravity but will fan out in a semicircle travelling sideways as well.

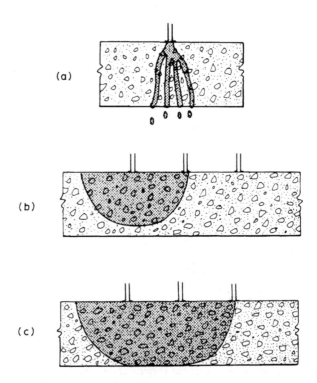

Fig. 4.8 Technique for containing resin to depth of member, or lesser depth if required. (a) Free-flowing resin simply drains out of crack by gravity. (b) Penetrative thixotropic resin is unaffected by gravity and fans out. Injection points spaced 5/6 of required depth of travel. Face of crack sealed between points. Stop pumping when resin reaches next point (use drinking straw as 'dip-stick'). (c) Repeat process from each point in succession. Resin will fill crack with minimal loss from unsealed (lower) face.

If there is no risk of resin escaping from the far side, the injection points may be more widely spaced and greater pressure may be used to make the resin travel further. If the far side is accessible injection points may be fitted there, but it is usually easier to do all the pumping from one side and only use the other side for 'witness points', closing them off when the resin begins to flow out from them. The spacing of points is very much a matter of experience with the particular resin system, the equipment available and an assessment of the crack (250 mm is an averge starting interval until you have the 'feel' of the job). It is always better to have too many points than not enough; when resin has been pumped from one point to the next, if it is still travelling well without the use of excessive pressure, the second point may be temporarily plugged (with a screw of paper towel) and pumping continued until the next is reached. When it becomes necessary to cease pumping from one point and move to another, the first point should be closed by crimping with long-handled pliers.

Another oft-recommended practice which the author has found totally unnecessary is to blow compressed air through the fitted injection tubes to remove dust from the crack before injecting resin. The small amount of loose dust likely to be present will easily be bound by the injected resin and will have no effect on the strength of the bond. However, if the crack has been subjected to heavy contamination by the ingress of water-borne dirt or by oil, some flushing may be necessary. In the former case, water flushing accompanied by some air injection to provide agitation will be effective. After thorough drainage, possibly air assisted, a water-tolerant resin should be used for injection. In the latter case the crack can be flushed with a suitable solvent, again with air injection, and thoroughly drained. The author has demonstrated this technique to be effective by achieving total structural rebonding when cracks heavily contaminated with dirty engine oil have been flushed and resin injected.

Injecting

As might be expected, the choice of pumping equipment depends upon the scale of operations. For small jobs a hand-held mastic gun is sufficient. Some formulators supply materials in disposable hand guns. In one type, a tube of hardener is screwed onto the resin-filled cartridge and the pair is inverted and reverted repeatedly for mixing before being fitted into a skeleton gun; in another, a twin-component-filled cartridge is fitted with a special mixing nozzle which blends the two components as they are pumped through simultaneously. With a non-disposable barrel gun the resin and hardener must be blended and poured into the gun; with the somewhat larger garage grease pumps, the intake of the pump must be lowered into the container.

The disadvantage with pumps using blended resin and hardener is the very limited pot life (typically 20 minutes) of the mixed material. The mix-in-the-nozzle guns have the advantage of keeping the components separate until they reach the nozzle and this is seen again in the larger scale equipment from which they were developed — the twin metering–mixing pumps. Reservoirs containing typically 25 litres each of resin and hardener feed twin volumetric pumps which supply the materials in the appropriate ratio to the mixing head. These are used for large-scale contracts.

There are some ill-founded fears about the effect of pressure injection upon cracked concrete. Although the actual line pressure may be 3–5 bar, sometimes increasing up to 8 bar for a tight crack, a little consideration will make it plain that as the resin fans out from the point of entry the pressure falls dramatically, to become negligible a few centimetres away. The author's experience has been that if the concrete is weak and vulnerable the cracks will be fairly wide and require very little pressure to fill, whereas fine cracks demanding higher pressures only occur in strong concrete that is well knit together.

There is little further to be said about the actual process of injecting the resin, as most points have already been discussed in the reviewing of crack preparation, the positioning of injection points and the choice of pump. It only remains to be said that when using low viscosity easy-flowing resins the injection process should commence at the lowest point, continue until clean resin free of air bubbles emerges from the next tube and then move progressively upwards. If there is water in the crack the first resin coming through will appear 'milky' and pumping should continue until clean uncontaminated resin appears. If a thixotropic resin system is being used (this is recommended in practically every case) then gravity has no effect and pumping may commence at the widest part of the crack working progressively to the finer extremities.

Although the mixed resin system may have a very short pot life in bulk in the gun, when it is injected into the crack there will be no build-up of exothermic heat and it may take some hours to harden, so it is best not to remove injection tubes until the following day.

Except when disposable equipment is used it will be necessary to clean out the equipment immediately after injection. Unused mixed resin should be pumped into a can and sand or gravel stirred in and left to harden for safe disposal. The equipment should then be dismantled and cleaned. The resin supplier will recommend the safest solvent for this, but it is of great advantage to have the solvent made into a thixotropic paste by the incorporation, with high shearing, of a few per cent Bentone. The solvent may then be used much more economically (it does not saturate the rags or tissues used for wiping out), the dangers of spillage are eliminated and nozzles and pipelines can be purged much more effectively; whereas free-flowing solvents will only bore a hole through a mass of resin, the thixotropic paste displaces it more completely. The forcing of polythene foam 'pigs' down the line by pumping solvent paste is the best way of purging them.

4.3.8 Sprayed concrete

Where damage due to reinforcement corrosion or fire has caused widespread spalling, the original section can be restored using sprayed concrete. Figure 4.9 shows typical details for the repair of a fire-damaged downstand beam.

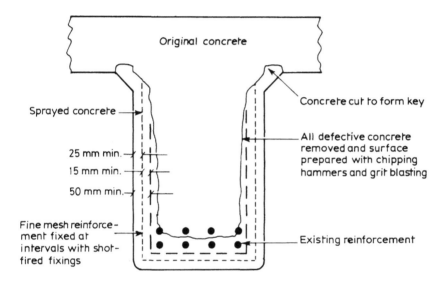

Fig. 4.9 Beam repair using sprayed concrete.

The cementitious material is referred to as gunite when the maximum aggregate size is less than 10 mm. For aggregates of 10 mm or greater the material is referred to as shotcrete. The material is sprayed into place at high velocity through a nozzle using either a wet or a dry process.

The dry process

The cement and aggregate are pre-mixed dry and fed into a delivery machine which in turn feeds the mixed material in a stream of high velocity air into a flexible delivery hose. Water is introduced as a spray at the discharge nozzle. Delivery hoses are typically 25–30 mm in diameter and can transport material up to 300 m horizontally or 100 m vertically.

Aggregate–cement ratios are typically 3.5 or 4 to 1 but aggregate rebound occurs on impact, resulting in a richer mix *in situ* than in the pre-mix. Rebound can represent between 5% and 30% of the material. Particular care has to be taken to ensure the proper filling of narrow faces and behind reinforcement bars. It is also important to ensure that the rebound does not become lodged in the work.

Given these precautions, a high density, low permeability placed material can result with strengths of 40–50 N/mm².

The wet process

Conventional pumping equipment is used to transport the wet pre-mix along a delivery hose to the discharge nozzle. Additional air is used to impart a velocity sufficient to achieve compaction on impact.

The water–cement ratio will be higher than in the dry process, although admixtures can be introduced to improve workability, accelerate setting or improve adhesion. Strengths tend to be lower (20–30 N/mm²), although problems associated with trapping of rebound and variability are less than with the dry process.

The success of both processes relies heavily upon the experience of the nozzleman. Application of the spray should be approximately at right angles to the surface while the optimum distance from the nozzle to the surface lies between 0.5 and 1.5 m.

Fibres can be incorporated into the mix to enhance performance. Steel wire fibres can improve flexural strength and abrasion resistance but must be overlain with a non-fibre cover layer to avoid unsightly corrosion. Alternatively, polypropylene fibres can be employed to improve impact resistance.

The Concrete Society has issued a Specification (1979) and Code of Practice (1980) for sprayed concrete, both of which make reference to the preferred methods of forming construction joints (Fig. 4.10).

The gunite process can benefit from the use of more sophisticated materials than simple sand–cement and accelerators. Many of the polymer-modified formulations originally developed for hand application have been successfully applied by spray using both wet and dry processes. The advantages are considerable: the use of more cohesive mixes, often of somewhat lower density, permits high build without the use of accelerators. This allows trowel finishing of the surface without damage or disturbance of the bond, whereas the normal flash-set gunite is not very amenable to finishing operations. Much greater build can be achieved with these materials without the need for supporting mesh reinforcement and, by suitable formulation, the mesh is not required for crack control. The benefits of improved mechanical properties and much greater resistance to the penetration of chlorides and carbon dioxide afforded by these materials has already been covered in the earlier sections of this chapter.

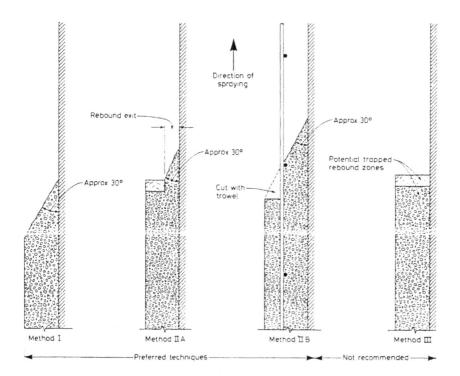

Fig. 4.10 Methods of forming construction joints in sprayed concrete (Concrete Society, 1979)

4.4. Safety

No discourse on the use of repair materials would be complete without referring to safe handling precautions. Safety aspects should partner the measures practised within a quality system and they represent a vital part of the education and training of the personnel involved. Legislation requires the material supplier to label and classify products (as flammable, irritant, harmful, corrosive or poisonous), and to include standard risk phrases and safety procedures. Product information sheets must also be supplied in accordance with the Health and Safety at Work Act, detailing a large range of product data and safe handling procedures.

Many resinous repair materials contain no solvents but these may be used for cleaning up operations. Strict control must therefore be exercised over such procedures. The common mechanical abrasion techniques used for removing material and for preparing surfaces (such as grinding, blasting etc.)

also carry an obvious, albeit limited, risk. On the whole, common-sense precautions such as the use of skin and eye protection are sufficient for many applications. For further information the reader is referred to the CIRIA publication on the safe use of chemicals in construction (CIRIA, 1981), the new Control of Substances Hazardous to Health (COSHH) legislation and, of course, the manufacturer's recommendations.

References

ASTM (1983) *Standard Test Method for Thermal Compatibility between Concrete and an epoxy resin overlay: ASTM C 884–78.* ASTM Annual Book of Standards.

BSI (British Standards Institution) (1971) BS 4652: *Metallic Zinc-rich Priming Paint*, BSI, London.

BSI (1978) BS 12: *Ordinary and Rapid-hardening Portland Cement*, BSI, London.

BSI (1983) BS 6319: *Testing of Resin Compositions for Use in Construction.* Part 1: *Method for Preparation of Test Specimens*, BSI, London.

BSI (1984) BS 6319 *Testing of resin compositions for use in construction. Part 4: Method for measurement of bond strength (slant shear method)*, BSI, London.

BSI (1984a) BS 1881: *Methods of Testing Concrete.* Part 105: *Method for Determination of Flow*, BSI, London.

Chung, H.W. (1975) Epoxy-repaired reinforced concrete beams, *ACI Journal, Proceedings*, **72** (5) 233–4.

CIRIA (Construction Industry Research and Information Association) (1981) *A Guide to the Safe Use of Chemicals in Construction*, CIRIA Publication SP16, London.

Concrete Society (1975) Polymer concrete. *Technical Report 9*, Concrete Society (draft revision, 1987).

Concrete Society (1979) *Specification for Sprayed Concrete*, Publication 53.029, Concrete Society.

Concrete Society (1980) *Code of Practice for Sprayed Concrete*, Publication 53.030, Concrete Society.

Concrete Society (1982) Non-structural cracks in concrete, Report of a Working Party, *Technical Report 22*, Concrete Society.

Cox, R.N., Coote, A.T. and Davies, H. (1989) Results of exposure tests to evaluate repairs to reinforced concrete in marine conditions, *Information Paper IP 18/89*, BRE, London.

Department of Transport (1986) *Materials for the Repair of Concrete Highway Structures*, Departmental Standard BD 27/86, Department of Transport, London.

Dhir, R.K., Hewlett, P.C. and Chan, Y.N. (1986), Near surface characteristics of concrete — an initial appraisal, *Magazine of Concrete Research*, **38** (134), 54–6.

Dixon, J.F. and Sunley, V.K. (1983) Use of bond coats in concrete repair, *Concrete*, **17** (8), 34–5.

Dubois, J. (1981) *Etude de l'Incidence des Injections dans les Fissures avant la Mise en Précontrainte*, CSTC revue no. 2, June.

Emberson, N.K. and Mays, G.C. (1990a) The significance of property mismatch in the patch repair of structural concrete: (1) Properties of repair systems, *Magazine of Concrete Research*, **42** (152) 147–60.

Emberson, N.K. and Mays, G.C. (1990b) The significance of property mismatch in the patch repair of structural concrete: (2) Axially loaded reinforced concrete members, *Magazine of Concrete Research*, **42** (152) 161–70.

Gledhill, R.A. and Kinloch, A.J. (1974) Environmental failure of structural adhesive joints, *Journal of Adhesion*, **6**, 315–30.

Hewlett, P.C. and Morgan, J.G.D. (1982) Static and cyclic response of reinforced concrete beams repaired by resin injection, *Magazine of Concrete Research*, **34** (118), 5–17.

Judge, A.I. and Lambe, R.W. (1987) Selection and performance of substrate priming systems for cementitious repairs, *Structural Faults & Repair '87*, Engineering Technics Press, Edinburgh, pp. 373–7.

McCurrich, L.H., Cheriton, L.W. and Little, D.R. (1985) Repair systems for preventing further corrosion in damaged reinforced concrete, *Proceedings of the 1st International Conference on Deterioration of Reinforced Concrete in the Arabian Gulf*, Construction Industry Research and Information Association (CIRIA), London.

Staynes, B.W. (1988) Delamination of polymer composite sandwich structures – the interaction of controlling parameters, *Proceedings of the 2nd International Symposium on Brittle Matrix Composites, Brittle Matrix Composites 2* (eds Brandt, A.M. and Marshall, I.M.), Elsevier Applied Science, London.

Tabor, L.J. (1985) *Effective Use of Epoxy and Polyester Resins in Civil Engineering Structures*, Construction Industry Research and Information Association (CIRIA), London.

Protection   5

K.G.C. Berkeley
(Corrosion Prevention Consultants)

A. CATHODIC PROTECTION

5.1 Introduction to cathodic protection

5.1.1 The background to corrosion attack

Cathodic protection is a technique for inhibiting corrosion which is becoming more widely considered where corrosion is a problem in existing reinforced concrete structures. There are three basic causes of corrosion on steel:

1. chemical attack;
2. atmospheric effects;
3. electrochemical effects.

Cathodic protection is **not** appropriate as the primary means of combating corrosion of structures affected by gaseous diffusion: a method involving some form of barrier separation would usually be more effective (section 5.6 *et seq.*). However, cathodic protection should be considered wherever moisture is present, particularly when it can react with chemicals either within the concrete or introduced from outside. Thus cathodic protection can be successfully applied to buried, immersed or above-ground structures.

The corrosion of steel-reinforced concrete is affected by both the chemical and the atmospheric environment; it is therefore important to understand the electrochemistry of the system, especially when recommending and applying remedial measures. Like many other scientific discoveries the first practical applications of electrochemical solutions to corrosion problems were introduced in the early Victorian period by Sir Humphrey Davy, following earlier investigations by Faraday. These ideas were taken up in the 1890s by, amongst others, Edison who attempted to apply the techniques to ships' hulls. This was followed by practical installations in various parts of the world, including Russia and the USA, until World War II. This was an era of considerable development, as for so many other scientific techniques. Probably the first full-scale use of cathodic protection in the UK was on the PLUTO network of cross-country oil pipelines, part of which is still in use today, followed by the Royal Navy hull systems of the early 1950s.

The application of cathodic protection to concrete has slowly grown since the mid-1950s, when it was successfully fitted to prestressed concrete pipelines, to the current systems on bridges, car park decks, storage tanks and buildings (National Association of Corrosion Engineers, 1985). The expanding use of this method has been primarily controlled by the availability of suitable materials for long-life anodes. These are only just starting to come on the market for above-ground structures, but experience is growing in the application of cathodic protection to reinforced concrete so that it can now receive serious consideration as a proven alternative rehabilitation method.

The principal use of cathodic protection at the present time is on structures subjected to chloride attack because it offers virtually the only alternative replacement. The earliest systems were fitted to control corrosion on the reinforcement in piling and dock gates immersed in seawater using variations on methods and materials used on steel jetties and pipelines. As new anode systems have become available these principles have been extended to structural reinforcement, e.g. where de-icing salts cause rapid corrosion of bridge decks.

As well as chlorides from external sources, it is now generally accepted that the chloride content of an aggregate is more critical than had previously been realized and standards permitted chloride contents which were too high for long-term safety. In fact, almost any acid radical in concrete has the potential to act as a corrosion initiator for the reinforcement. Where these radicals penetrate to the steel cage, corrosion can occur to a greater or lesser extent when moisture is present. A slightly different corrosion mechanism can occur with carbonation which alters the chemistry of the concrete and so exposes the reinforcement to atmospheric attack from outside.

Whenever a standard survey procedure on a reinforced structure indicates possible high chloride contents then the probabilities of short- or long-term reinforcement degradation arises, and both potential mapping and resistivity surveys should be carried out as described in Chapter 3.

5.1.2 Electrochemical corrosion

Corrosion, other than that arising from gaseous diffusion, is a process which requires an electrical potential difference to exist between different sites on the surface of the material and that these be connected by an electrolyte. If a piece of steel is immersed in water (Fig. 5.1) it will corrode because of the inhomogeneity of the surface microstructure in contact with the water. The corrosion process involves dissolution of the steel at the anode (positively charged) whereby ferrous ions move in solution to the negatively charged part of the surface, the cathode. This ionic movement creates an electron

Alternatively charged surface microstructure

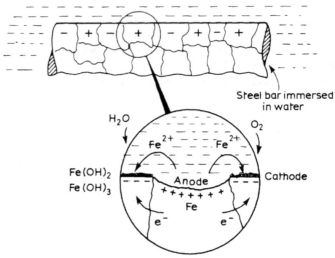

Fig. 5.1 How corrosion occurs:

Anodic reaction	*Cathodic reaction*

$Fe \rightarrow Fe^{2+} + 2e^-$

$Fe + 2H_2O \rightarrow Fe(OH_2) + 2H^+ + 2e^-$
ferrous hydroxide

$2Fe + 3H_2O \rightarrow 2Fe(OH)_3 + 3H^+ + 3e^-$
ferric hydroxide

$3Fe + 4H_2O \rightarrow Fe_3O_4 + 8H^+ + 8e^-$
magnetite

$4e^- + O_2 + 2H_2O \rightarrow 4(OH)^-$

$2H^+ + 2e^- \rightarrow H_2$

movement and hence a current flow I. The role of chloride in inducing reinforcement corrosion in concrete is complex and not, as yet, fully understood. However, it is believed that chloride contained within the pore solution in concrete is able to enhance the dissolution of steel by disruption of the passive film.

The rate and extent of corrosion are primarily governed by Faraday's law and Ohm's law respectively. The current flow is controlled by the voltage difference created between sections of the surface and the resistance which it has to overcome to flow from one section to another, expressed as Ohm's law:

$$\text{voltage } V = \text{resistance } R \times \text{current } I$$

Resistance R is usually expressed in terms of resistivity, which is the electrical resistance of an electrolyte per unit length across a unit cross-sectional area. Values range from 20 ohm cm for seawater to 100 000 ohm cm for granite rocks. Related figures for concrete largely depend on the density and composition of the original aggregate and the environment to which it is exposed — varying from piling in seawater to structures above ground in dry climates. It is now recognized that the presence of moisture and chloride even in above-ground structures can make reinforced concrete a very active medium, lowering its resistivity to below 1000 ohm cm, in which active corrosion of steel reinforcement can occur (section 3.3.1).

Instrumentation is now available to test the resistivity of concrete on site (Fig. 5.2) and to measure the electrochemical state of reinforcing steel (Fig. 5.3). The anodic (corroding) and cathodic (protected) areas of a structure can be assessed by measuring resistivity and, in conjunction with chemical analyses, the aggressivity of the conditions to which the structural concrete has been exposed can be deduced. Techniques for quantification of metal loss over specific sections of reinforcement have not yet been fully developed

Fig. 5.2 A resistivity meter developed for concrete.

Fig. 5.3 Potential measurement and logging (courtesy of CMT Ltd, Raynesbury, Derby)

but it is expected that these will come with time. Survey engineers who are experienced at these newer techniques can make reasonably accurate judgements about present and future corrosion rates given a specific set of circumstances.

5.2 How cathodic protection works

Figure 5.4(a) translates the intergranular corrosion cells arising from the surface inhomogeneities referred to above and in Fig. 5.1 into a piece of corroding reinforcement set in concrete. It becomes apparent that, if these locally created cells can be neutralized so that the current is unable to flow, the corrosion reactions will not occur.

One way of achieving this is by totally coating the reinforcement bars, thereby electrically isolating them from the concrete. In new construction the use of epoxy-coated reinforcement bars is now becoming an accepted method but such an approach is not possible on existing structures. It is also sometimes possible to inject an inert chemical which sets in the concrete pores and so

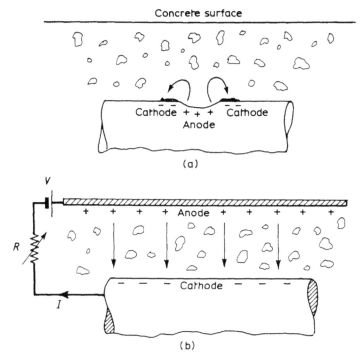

Fig. 5.4 (a) A freely corroding reinforcement bar (see Fig. 5.1 for details). (b) A cathodically forced surface.

limits moisture and chloride ingress, but this approach is relatively expensive and somewhat hit or miss unless extensive exploratory work is carried out.

It is possible to protect steel reinforcement from corrosion in an ionically conducting environment by connecting it to a current source which will supply electrons to the metal so that it becomes the negative side of the circuit, as shown in Fig. 5.4(b). The current can be gradually increased until all the small local cells are progressively reversed to the ultimate state of inhibition (Berkeley and Pathmanaban, 1989). This is the process known as 'cathodic protection' and, because of the very small current densities necessary, a relatively large atmospherically exposed structure can be treated using only a few hundred watts (watts = volts × amps) — typically 200 W which is equivalent to a large light bulb.

This external current can be provided in two ways: either by connecting it to a metal which is higher in the electrochemical series of metals than steel, e.g. magnesium, aluminium or zinc, or by applying a direct current to the reinforcement. The latter is usually transformed and rectified from a normal domestic alternating current supply. The choice of method generally depends on the environmental conditions of the structure and the voltage necessary to overcome its resistivity.

Cathodic protection can be applied to steel reinforcement in two types of concrete structure — those exposed to the atmosphere and those submerged or buried — and different solutions will usually be appropriate. The environmental parameters are totally different in the two situations. In the latter instance, if the reinforcement is corroding, the concrete is wet and probably chemically contaminated, e.g. with seawater. The resistivity is likely to be very much lower than with atmospherically exposed structures and the limited voltage outputs from dissimilar metal anodes can be economically exploited. It is only necessary to arrange a connection between the chosen metal and the steelwork for current to flow. The anode will then slowly dissolve over a period of months or years — this is termed 'sacrificial protection' (Fig. 5.5).

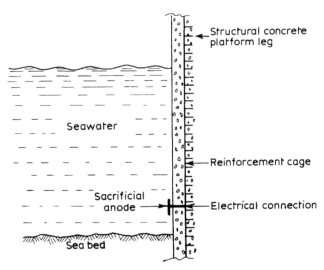

Fig. 5.5 Sacrificial anode(s) on a platform leg.

The resistivity of atmospherically exposed structures, even if they are relatively heavily contaminated with salts, is such that a higher voltage is required to spread the protection current around the whole reinforcement cage. The material through which the protective current is passed into the structure, and which becomes the positive half of the circuit (the anode), is made of an inert material which does not corrode with time. This current is supplied from a power unit not unlike a commercial battery charger, usually with a series of adjustable outputs. Each negative connection is to a section of the reinforcement cage as shown on the left-hand side of Fig. 5.4(b). These anodes are held onto the concrete surface by non-metallic fixings and oversprayed with gunite, or a suitable protective layer or coating, to ensure electrical conductance to the original structure and to provide a degree of protection (Fig. 5.6). This is called impressed current cathodic protection.

Testing for the degree of inhibition achieved by cathodic protection has traditionally been difficult, primarily owing to the fact that it is not possible to 'see' the conditions at the steel surface when it is buried under 50–100 mm of concrete. However, coupons and embedded electrodes are now available for this purpose. On one criterion, if the natural (or 'instantaneous off')

Fig. 5.6 Fixed anode cover coat on bridge column (courtesy of Raychem Ltd, Swindon, Wilts.).

potential value is made 200–250 mV more negative (with reference to a silver/silver chloride half-cell placed on the concrete surface) when the current is switched on, then the steel immediately below is probably protected.For a description of the role of half-cells in determining the potential of the steel see Chapter 3. 'Instantaneous off' is defined as the potential obtained not less than 0.1 s and not more than 1.0 s following interruption of direct current power to the anode system. On another criterion, the potential of the reinforcement is depressed to −770 mV but if these readings are taken on the surface of the concrete the gap of 50–100 mm to the reinforcement may introduce errors. The higher the natural moisture content of the concrete the more reliable are the readings. The Concrete Society (1989) recommends the following criteria for protection: a minimum of 100 mV potential decay over all representative points within the area of the structure being protected, subject to a most negative limit of −1.10 V with respect to a silver/silver chloride half-cell 'instantaneous off' potential.

5.3 Associated problems

Passing a current through the concrete will cause ionic movement whereby the negatively charged ions will be attracted to the anode and the positively charged ions to the cathodic steel surface (Fig.5.7). Whilst this process assists in restoring passive alkaline conditions to the reinforcement, care must be taken to ensure that overprotection does not occur, resulting in hydrogen gasification at the concrete–steel interface and thereby weakening the mechanical bond. The movement of acidic ions (chlorides, sulphates etc.) away from the steel is highly desirable as it removes the contaminants from the steel surface and progressively decreases the rate of corrosion attack. Conversely, in so doing, the pH of the anode area is decreased quite

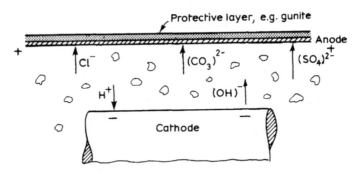

Fig. 5.7 The basic ionic movement.

considerably and, if not carefully controlled, delamination of the surface-mounted anode and cover may be initiated. Because these negatively charged ions, which include chloride, are moved to the anode there is a theoretical possibility that continual washing of the anode surface can progressively remove this chemical contamination in certain circumstances and so keep the surface pH within acceptable limits. The fact remains that whatever cathodic protection system is applied it has to be carefully checked at installation, and kept under a regular maintenance schedule thereafter, to maximize the inhibitive control and prevent unwanted side effects building up.

There are two other concrete states which are not well recognized and yet are the root cause of many problems. The first relates to the variation in oxygen content along a steel bar that gives rise to a basic 'differential oxygenation' cell. It can arise from a simple variation in chemical content of the aggregate or climatic conditions from one day to another. This can give rise to a different pore structure after curing which results in a different oxygen availability at the reinforcement at adjacent sections. A similar state can easily arise from patch repairing: usually the cut-out part is replaced by a mortar or compound mix which is less permeable and denser than the original concrete. This will alter the oxygen availability between the replaced section and the existing section which results in an increased corrosion rate in areas immediately adjacent to the cut-out but in the original concrete. It is important to recognize that this problem can only be overcome by absolute assurance that the cut back is into totally sound material — which may be uneconomical to achieve (Chapter 4).

Another corrosion cell which can be created in submerged concrete structures is caused by the natural potential difference between the reinforcing steel embedded in concrete and an associated piece of steel immersed in water or buried in soil when they are allowed to make a structural or fortuitous electrical contact (Fig. 5.8). The potential difference can be as high as 1000 mV and thus the voltage created is equivalent to that of a sacrificial anode with the section outside the concrete dissolving quite rapidly. A fairly common example is that of an access ladder inside a concrete reservoir where the ladder fixing brackets are either part of, or touch, the reinforcement of the tank. Similarly, although not below sea level, there is sometimes sufficient water retention on appendages such as mooring bollards on jetties, where they are almost certainly part of the structural steel network, to cause premature failure by corrosion of the fixing bolts.

Corrosion can also be induced by what are termed 'stray currents'. These can be picked up from 275/400 kV alternating current transmission lines, in which case the iron–rust interface acts as a crude rectification circuit, or from direct current traction systems, high amperage welding generators or nearby cathodic protection schemes. In addition very big concrete structures, such as offshore oil production platforms, can be affected by earth currents.

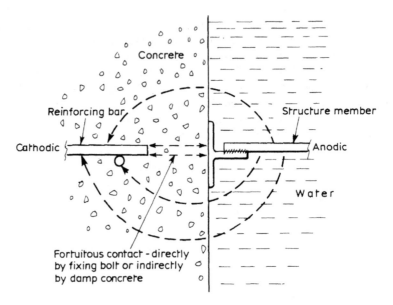

Fig. 5.8 A created corrosion cell in concrete.

In principle, where current is picked up from one of these sources the steel system balances itself by discharging the same amount of current at another, and probably remote, point. Within reason current pick-up is acceptable because it alters the steel cathodically, but at the discharge area it dissolves the structural steel at the rate of 9 kg per amp per year which on thin prestressing wires can be disastrous (Fig. 5.9). Even on 50 mm reinforcement bar if this happens over a relatively short time premature failure can be rapidly induced. For this reason alone all reinforcement cages on a structure should be bonded together in any areas of suspected stray current and a special earthing system should be considered. Such conditions have arisen on Metro tunnels being constructed of reinforced concrete sections where electrical tracton is envisaged (Kish, 1981) and on reinforced bridges close to high tension power transmission lines.

5.4 Practical applications of cathodic protection

As mentioned previously one of the earliest applications of this form of corrosion protection to a concrete construction was on prestressed pipe. The steel tensioning wire is wound over the spun concrete moulding and then

Fig. 5.9 Stray current attack causing premature failure of prestressing
wires on concrete pipe.

sprayed with a few millimetres of concrete as cover protection. Experience
showed that there were ground conditions (acidic, low resistivity) which
resulted in premature failure of the pipe under pressure and so very basic
sacrificial anodes were attached. Zinc was chosen as a suitable material because
it was readily available and cheap, and by chance this was one of the most
successful applications of this type of underground anode. One anode was
fitted per pipe length (3 m) and it proved to be a case of 'fit and forget'. In
later systems the significance of bonding all the pipes in one continuous length
was appreciated and this contributed to the use of more sophisticated cathodic
protection schemes, although some of these ran into control problems related
to the lack of maintenance.

A more up to date application of sacrificial anodes in a seawater environment is the system applied to some concrete oil platforms in the mid-1970s.
The importance of stray current pick-up was realized and a special bonding
circuit was added to the reinforcement to reduce the electrical resistance of
the total circuit. This bonding circuit was attached to a mass of anodes sited
just off the skirt at the base of the column so that they were always immersed

(Fig. 5.5). Stray current from any source was then transmitted along the bonding circuit, as it was the line of least resistance, and earthed to the anodes, which in turn were anodic to the steel of the structure (Heuze, 1980).

Because of the voltage necessary to spread the protection current around a normal reinforcement cage the impressed current system is much more widely used. Several systems are commercially available and the choice is related to initial cost, life expectancy and maintenance costs.

The two methods used are based on fixed anodes and conducting paints. The paint systems are carbon loaded (i.e. they contain a graphite conductive filler). They have an optimistic life expectancy of around 10–15 years and may be expensive to maintain, although this depends on the operating parameters and

Fig. 5.10 Metal oxide ('chicken wire') mesh fixed to bridge column (courtesy of ICI Chemicals & Polymers, Runcorn, Cheshire).

environmental conditions. One useful advantage is that they add little to the weight of a structure whereas the fixed anode systems are usually considerably heavier as they have to be coated in sprayed concrete/gunite (Fig. 5.6).

The choice of fixed anode systems is between the following.

1. a carbon-loaded polymer rod with central copper conductor core (Fig. 5.5);
2. a 'chicken wire' type metal oxide mesh (Fig. 5.10);
3. a ceramic 'stick-on' plate (Fig. 5.11).

Fig. 5.11 Ceramic plates fixed to bridge beam (courtesy of Ebonex Technologies Inc., California, USA).

Whatever system is used the preparations needed before applying a cathodic protection system are similar. The structural survey will have determined the areas requiring treatment which, in the case of the fixed anode system, will involve larger rather than smaller isolated areas of the structure. Cathodic protection is probably uneconomic for corrosion control on a surface area less than 10 m² but there is no upper limit to its effective coverage. Conducting paint is probably more flexible with respect to isolated or confined areas.

The first requirement before fixing a system to a structure is to remove all the spalled or loose concrete by whatever method is deemed appropriate in the circumstances. Any loose rust on the reinforcement is removed at the same time but there is no need to clean to bright metal. Reference electrodes should be fitted to allow for future testing to ascertain the efficiency of the system. The patch is reprofiled with cementitious grout, although the composition of this need not be to the original specification unless structural weakness is suspected (Chapter 4). The anode is then pinned to the structure in the case of fixed anodes, or dissipator wires are fitted prior to applying a conductive coating. Connector cables for the anode, reference electrode and the reinforcement cage are then run back to the transformer rectifier which is located at some suitable point usually not more than 25 m away (Fig. 5.12). These cables are protected and the whole system is either oversprayed with a suitable cement-based material in the case of the fixed anodes or with a cosmetic coat in the case of conducting paint. Ceramic anode systems are unusual in that each plate is fitted directly to the structure with the whole installation being interconnected by exposed cable (Fig. 5.11). They have the very useful

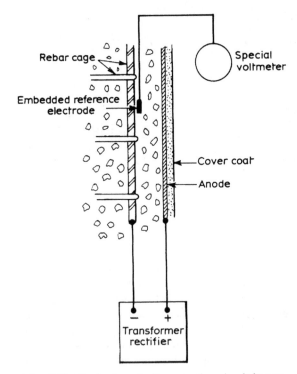

Fig. 5.12 The basic cathodic protection circuit layout.

advantage that they can work at a very much higher current output than any other system.

Once the power unit, which is the size of a commercial battery charger, has been energized the system can be commissioned and the circuits balanced to ensure an acceptable level of protection. As mentioned previously, a universal testing technique has not yet been agreed because of the wide variety of constructions and environments in which concrete structures are employed. For instance, the level of protection appropriate for reinforced concrete in seawater is not necessarily suitable for atmospherically exposed structures.

Attention is drawn to the comprehensive report, 'Cathodic Protection of Reinforced Concrete', prepared by a group of engineers interested in corrosion problems related to atmospherically exposed reinforced concrete, which offers useful guidelines and references on the application of cathodic protection to such structures (Concrete Society, 1989).

5.5 Concluding remarks

The problems of chloride-induced corrosion in reinforced concrete are only just beginning to be quantified. For instance, about half of the reinforced concrete bridge decks in the USA, as well as innumerable parking decks, are suffering from severe reinforcement corrosion. According to Leach *et al.* (1987), corrosion accounts for just over 4% of the gross national product of industrialized countries and so this problem cannot be ignored.

Both polluted atmospheric conditions and the chemical constituents of seawater, in particular chlorides, tend to reduce the protective pH of concrete around a steel bar. No concrete repair or remedial method can put corroded metal back onto the reinforcement, but on any structure where degradation is visible or is anticipated cathodic protection is the only method capable of directly reversing the physical phenomena causing the problem in the first instance. It is particularly suitable for large structures where massive patch repair costs appear to be the only alternative to demolition and rebuilding. When correctly surveyed, fitted and maintained, cathodic protection, in many instances, is well under half the cost of other remedial approaches. As such it must attract much greater attention from structural engineers, surveyors and architects in the years to come.

J.G. Keer
(Formerly University of Surrey)

B. SURFACE TREATMENTS

5.6 Introduction to surface treatments

Surface treatments are applied to a concrete structure to improve its appearance or to protect the structure from potentially aggressive agents. The protective benefit is considerd here, the requirements being that the surface treatment shall either enhance the durability of a new structure or extend the service life of an existing one. Methods for improving the surface properties of concrete by treatments have been investigated for many years (Washa, 1933) and many standard texts on concrete technology have included sections on surface treatments. Past emphasis has perhaps been on protection of concrete against chemical attack when concrete is used in non-typical rather special situations. For example, the report of the American Concrete Institute (ACI) Committee 515 (ACI, 1966), first published in 1966 as a 'Guide to the protection of concrete against chemical attack' and subsequently revised to cover various barrier systems (ACI, 1985), contains a wealth of information on different protective systems against over 250 potentially aggressive agents from acetic acid to zinc sulphate.

Recent attention, however, has focused on the use of surface treatments to improve the performance of concrete structures in typical environments and applications which form the 'regular' use of concrete. The principal reason has been the desire to reduce the occurrence of reinforcement corrosion and associated damage. The interest in surface treatments has shifted to their role in slowing down the ingress of carbonation or chlorides, which can initiate corrosion of embedded reinforcement, as described in Chapter 2. This role is therefore the focus of this part of the chapter, together with brief discussions on the protection against freeze–thaw damage and alkali–silica reaction (ASR). It should be noted throughout that the success of surface treatments depends to a great extent on the standard of surface preparation of the substrate.

5.7 Types of surface treatment

One way of classifying the large number of types of surface treatment available for concrete is by the manner in which the protection or benefit is achieved. There are four main classes of treatment: coatings and sealers, pore-liners, pore-blockers and renderings. The distinction between them is shown diagrammatically in Fig. 5.13. Table 5.1, based on a classification of this form by the Construction Industry Research and Information Association (CIRIA, 1987), indicates the more common materials within these broad classes together with the form of the fresh treatment, its method of cure and colour.

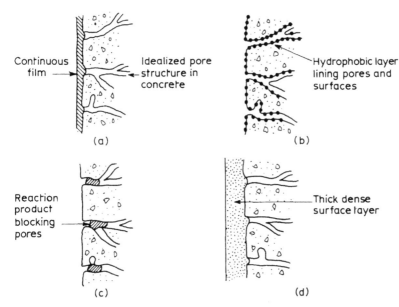

Fig. 5.13 Classes of surface treatment: (a) coating/sealers; (b) pore-lining treatment; (c) pore-blocking treatment; (d) rendering.

5.7.1 Coatings and sealers

Coatings and sealers rely upon the formation of a pinhole-free film of finite thickness over the concrete surface that acts as a barrier to the passage of water or the diffusion of carbon dioxide, for example. Those materials described as sealers claim some penetration into the pores of the concrete. Most are what would commonly be called paints and may be pigmented to provide colour. Coatings normally have a dry film thickness of 100–300 μm, applied in two or more layers, although thicker 'high-build' coatings are also used.

Table 5.1 Classification f surface treatments (CIRIA, 1987)

Classification material		Form of liquid paint	Cure	Colour
Coating sealer	Acrylic	Catalysed, solvented (sealer)	Loss of solvent and chemical	Clear
		Solvented	Loss of solvent	Clear and Pigmented
		Aqueous dispersion	Drying	Pigmented
		Cementitious aqueous dispersion	Drying	Pigmented
	Butadiene copolymer	Aqueous dispersion	Drying	Pigmented
	Chlorinated rubber	Solvented	Loss of solvent	Pigmented
	Epoxy resin	Catalysed, solvented sealer	Loss of solvent and chemical	Clear and pigmented
		Catalysed, aqueous dispersion	Drying and chemical	Clear and pigmented
		Catalysed, solventless	Chemical	Pigmented
	Oleoresinous	Solvented	Loss of solvent and oxidation	Pigmented
	Polyester resin	Catalysed, solvented (sealer)	Loss of solvent and chemical	Clear and pigmented
	Polyethylene copolymer	Solvented	Loss of solvent	Pigmented
		Aqueous dispersion	Drying	Pigmented
	Polyurethane	Solvented (sealer)	Loss of solvent and moisture	Clear and pigmented
		Catalysed, solvented (sealer)	Chemical	Clear
		Catalysed	Chemical	Pigmented
	Vinyl	Solvented sealer	Loss of solvent	Clear
		Aqueous dispersion	Drying	Pigmented
Hydrophobic Pore liner	Silicones	Solvented	Loss of solvent and reaction	Clear
	Siloxane	Solvented/solventless	Loss of solvent and reaction	Clear
	Silane	Solvented/solventless	Loss of solvent and reaction	Clear
Pore blocker	Silicate	Aqueous dispersion or solvented	Reaction	Clear
	Silicofluoride	Aqueous dispersion or solvented	Reaction	Clear
	Crystal growth materials	In cementitious slurry	Reaction	Cementitious
Rendering	Plain and polymer modified cement-based mortars	—	Cement hydration	Cementitious

Coatings and sealers are composed of a number of constituents. The basic film-forming material is the binder and it is by the composition of the binder that generic types of coating are classified. Thus coatings and sealers used for concrete include epoxies, polyesters, acrylics, polyurethanes, vinyls, butadiene and polyethylene copolymers, alkyds, bitumens and oleoresinous varnishes such as linseed oil. The fluid content of the binder can be increased and the viscosity reduced by a solvent or diluent. The latter term is applied to liquids that do not dissolve the polymeric binder.

Various additives may be included to act, for example, as plasticizers, catalysts or fungicides. Fillers can be blended into the coating to allow it to be applied at relatively great film thicknesses or to improve abrasion resistance. Pigments, finely ground opaque powders, are used to colour coatings. Thus, as well as the availability of a number of basic film-forming binders, many formulations are possible based on a single binder, and relatively minor variations in formulation can lead to very substantial changes in performance. It is recognized, for instance, that the volume concentration of the pigment can have a significant effect on anti-carbonation performance (section 5.9).

The hardened surface film may form from the liquid application in a number of ways: evaporation of the solvent or dispersant (water) as with an acrylic emulsion; chemical reaction stimulated by a hardener or catalyst as with epoxy-resin-based materials; reaction with moisture in the atmosphere or substrate as with a one-pack polyurethane formulation; reaction with oxygen (oxidation) in the atmosphere as with an oleoresinous coating such as linseed oil (Table 5.1). One important property of coatings for concrete is that the integrity of the film should not be destroyed by the alkalinity of the concrete substrate, i.e. the coating should be non-saponifiable. Oil-based paints, for example, have poor alkali resistance and are used with alkali-resistant primers or in special alkali-resistant formulations.

For a fuller description of the composition of types of coating or sealer, the reader is referred to more specialized texts (CIRIA, 1987; Turner, 1988).

5.7.2. Pore-lining treatments

Pore-lining treatments involve hydrophobic materials which line surface pores of the concrete and repel moisture by the mechanism indicated in Fig. 5.14 (Edmeades and Hewlett, 1985). The most important materials in this class are based on silicone compounds, which are synthetic chemicals produced by the chemical combination of inorganic materials containing silicon and oxygen with hydrocarbons. Figure 5.15 summarizes the mode of action of the main compounds in the silicone family of treatments (Roth, 1986).

Silicone resin itself has been used as a concrete and masonry treatment for many years. For application the silicone resin is dissolved in an organic

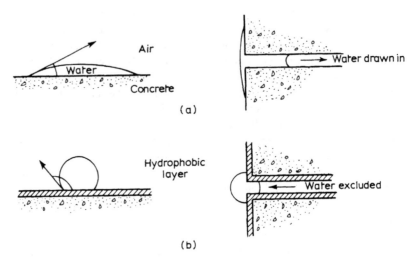

Fig. 5.14 Effects of a hydrophobic pore-lining surface treatent (Edmeades and
Hewlett, 1985): (a) ordinary concrete, contact angle less than 90°;
(b) concrete with water-repellant surface, contact angle greater than 90°.

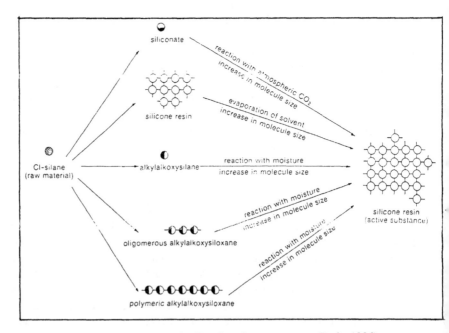

Fig. 5.15 The silicone family of surface treatments (Roth, 1986).

solvent (at approximately 5% by weight) which, on evaporation, deposits the water-repellent silicone resin on the walls of the concrete pore structure. The silicone can link chemically with a cementitious substrate.

The silane materials, which may be used undiluted or diluted in, typically, ethanol or white spirit, react with moisture present within the pore structure to form silicone resin, chemically linked to the substrate. The reaction is not spontaneous and probably takes place over a period of days. Catalysts may be incorporated in the treatment to accelerate the reaction, the speed of which is also related to the pH of the pore water. The hydrolysis reaction occurs more quickly at higher pH. At neutral values of pH the reaction time can be very prolonged (Robinson, 1987a). The alkylalkoxysilanes have much smaller molecules (molecular dimension about 10 angstroms) than the silicone resin which eventually forms and can penetrate further into the concrete (typically 2–3 mm) than a directly applied silicone. A more precise definition of the silane type specifies the composition of the alkyl and alkoxy groups — hence treatments may be based, with increasing molecular weight, on methyltrimethoxysilane, isobutyltrimethoxysilane or isooctyltrimethoxysilane. While the low molecular weight silanes might be expected to penetrate into concrete more easily, they are volatile substances and a proportion of the applied silane can evaporate. Oligomeric siloxanes are formed from the polymerization of the monomeric silane unit into short-chain molecules; polymeric siloxanes combine more units into longer chains. The advantage of these compounds is that they are much less volatile than the monomeric silanes. Solutions made up of the less volatile of the monomeric silanes (such as isooctyltrimethoxysilane) and an oligomeric siloxane have therefore been developed to produce a less volatile surface treatment.

Greatest experience with silane treatments has been obtained in West Germany and the USA, although applications in the UK to new structures and repaired structures are increasing.

Water-repellant treatments based on solutions of complex stearates (approximately 8% solids in white spirit) are also used. Like silicone resins, they were primarily developed as masonry treatments and their performance is similar to that of the silicone resins.

5.7.3 Pore-blocking treatments

Pore-blocking treatments make use of a family of products which are claimed to penetrate concrete and block pores. For some materials the pore-blocker is a product from the reaction between the penetrant and compounds already present in the concrete. The most common examples in this group are liquid silicates and liquid silicofluoride applications, which react with calcium hydroxide in the concrete to form, for the silicate application, further C–S–H gel products and, for the silicofluoride, insoluble calcium silicofluoride. Some sealers, such as epoxy or acrylic resins, may have sufficient penetration in special formulations to be considered as pore-blockers on hardening within the substrate by chemical reaction. Penetration in these cases can be improved by vacuum impregnation.

The principal application for pore-blocking treatments for many years has been as surface hardeners to prevent the dusting of concrete floors (Shaw, 1980) and to improve the abrasion resistance of concrete roads, although more recently liquid silicates and pore-blocking slurry treatments (see below) have been used as part of repair and protection systems for concrete damaged by reinforcement corrosion (Fletcher, 1982). Some treatments applied as cementitious slurries to a concrete surface contain active constituents which are claimed to migrate from the slurry coating into the concrete substrate. Pore-blocking reaction products form so that, if the slurry rendering spalls away, the parent concrete remains enhanced. These treatments are sometimes termed crystal growth methods, since the crystalline reaction products are able to bind with water molecules and increase in volume. The consumption of calcium hydroxide may also allow further cement hydration (Anon, 1978).

5.7.4 Renderings

A rendering is a thick cementitious coating, generally applied by trowel or by spraying as in the gunite process, which provides an additional barrier to the concrete cover against the ingress of potentially aggressive agents. Cement mortars can be modified by polymer latices, for example, to provide a relatively dense impermeable barrier and to enhance adhesion to the substrate. The surface renderings of the pore-blocking crystal growth materials mentioned above may be as beneficial in providing added protection as the pore-blocking of the substrate achieved by migration of active chemicals.

5.8 Application of surface treatments

Correct application is vital to the success of surface treatments. Detailed guidance is available in a number of publications, including BS 6150 (1982) for the painting of buildings and Building Research Establishment *Digest 197 and 198* (BRE, 1977). Surfaces should be sound, clean and free from dust, grease, mould oil, curing membranes, laitance, fungi, algae and lichens. For coatings and sealers, concrete should be dry. If moisture is picked up by absorbent paper pressed onto the face of the concrete, then the surface is likely to be too damp for most surface treatments.

Treatments of the silane type can be applied to damp concrete, but better penetration is achieved with dry concrete, where there is still sufficient moisture present in the pores for the hydrolysis reaction which converts the silane into silicone resin. The Department of Transport Standard BD 27/86 (1986) covers silane applications to repaired areas and requires two applications, each at a rate of 300 ml/m², to surface-dry concrete protected from the weather before application if necessary. Renderings and pore-blocking materials normally require the substrate to be damp. Manufacturers of treatment systems generally make recommendations regarding the desirable moisture content of the substrate.

To obtain a clean and mechanically sound surface, proper surface preparation is essential. Of the methods available, wire brushing is perhaps the least effective, bush hammering can be too vigorous and needle gunning, grit blasting or water blasting are likely to be the most effective. Filling of small surface voids and general levelling of the surface may be necessary. Very high or low temperatures and humidities during application will adversely affect most treatments. Coatings should normally be applied in a minimum of two coats to achieve a pinhole-free film. Silane type treatments are usually applied by low pressure spraying and saturation flooding of horizontal surfaces is desirable. With each type of treatment, manufacturers' recommendations for application should be followed carefully. Good application, skilfully combined with careful supervision, followed by inspection and checking are required to ensure correct coverage.

The useful service life of a surface treatment for concrete will depend on the type and formulation of the treatment, but it appears that, correctly applied, service lives of about ten years are achievable. For most structures this implies a continuing maintenance commitment and retreatment of surfaces. For coatings, it may be desirable and acceptable to overcoat the existing treatment (provided adhesion is still good) rather than clean it off completely. Only a few paints are incompatible with others during application, e.g. epoxy or chlorinated rubber paints on oil-based paint, but the adhesion between very different types can be poor. It is safer to continue the use of a coating type which has proved satisfactory. When a change has to be made for some reason, a trial patch and/or specialist advice may avoid a costly failure.

5.9 Surface treatments to control carbonation

The treatments used to control the ingress of carbonation are principally drawn from the coating/sealer class of materials. The role of the surface film is to limit the diffusion of carbon dioxide through it to a low value. The relative effectiveness of such anti-carbonation coatings has been assessed in three ways.

1. measurement of the carbonation of concrete coated with the treatment and exposed in natural conditions (Hankins, 1985);
2. measurement of the carbonation of mortar blocks coated with the treatment

Table 5.2 Comparison of the equivalent air layer thicknesses R of a number of coating systems (Robinson, 1987)

Coating system	Initial dry film thickness (μm)	Equivalent air layer thickness R (m) after t hours of accelerated weathering				
		t = 0	t = 500	t = 1000	t = 2000	t = 3000
W-b emulsion, 80% solids heavily filled with TiO_2 and mica	300	115	417	482	—	—
As above with low viscosity methacrylate/siloxane pretreatment, unp	300	150	210	127	—	—
Moisture curing urethane, p, s, heavily filled	170	71	289	253	383	656
Methyl methacrylate, s, p, externally plasticized, 18% PVC	200	42	150	332	283	—
Urethane, two-pack, D-D, s, p	110	216	281	505	221	—
Acrylic, w-b, translucent, low PVC	200	367	208	168	247	—
Styrene acrylate, p, 65.2% PVC, hb, s	240	310	—	72	165	—
Styrene acrylate, unp, w-b	160	115	14	1.5	—	—
Styrene acrylate, unp, w-b	150	53	21	1	—	—

Styrene acrylate, p, 57% PVC, s	150	100	30	75	90	—
Styrene acrylate, p, 14% PVC, s	100	61	75	134	268	—
Methylmethacrylate, p, 30% PVC, internally plasticized, s	140	143	379	363	—	—
Ethylene copolymer, w-b, p, 21% PVC + solvented acrylic pretreatment	300	436	57	153	447	—
Methylmethacrylate, p, 43% PVC, s	140	262	158	316	364	—
Acrylic, w-b, p	200	385	90	266	461	—
Styrene acrylate, p, s, hb	900	15	54	34	—	—
Alkyd, hb, s, + low viscosity sealer coat	1000	57	10	19	45	—
Acrylic, w-b, p, fibre reinforced + epoxy w-b pretreatment	200	145	150	298	—	—
Chlorinated rubber, hb, p, s	200	428	384	298	—	—

p, pigmented; unp, unpigmented; s, solvented; w-b, water-based; hb, high build; PVC, pigmented volume concentration.

and exposed to carbon dioxide in a test cabinet (Jones, 1983; McCurrich and Whitaker, 1985; Moller, 1985);

3. measurement of carbon dioxide transmission through a film of the coating supported on a paper, card, glass or ceramic substrate (Sherwood and Haines, 1975; Klopfer, 1978; Desor, 1985; Engelfried, 1985; Robinson, 1986, 1987b).

Data from type 1 tests are clearly most realistic; types 2 and 3 are accelerated tests providing rapid results useful for comparative purposes.

Tests of type 3 allow the calculation of the diffusion resistance coefficient μ_{CO_2} for carbon dioxide, a dimensionless parameter indicating how many times more resistant a coating is than air under equivalent conditions (Klopfer, 1978). The diffusion equivalent air layer thickness R in metres is obtained from the product of μ and the dry film thickness S of the coating (in metres), i.e.

$$R = \mu S$$

The higher the value of μ or R, the greater is the resistance of the coating to carbon dioxide diffusion through it. Comparison of the μ_{CO_2} or R values allows the relative effectiveness of different anti-carbonation coatings to be assessed. Table 5.2 (Robinson, 1987b) is typical of the data available in the open literature. Initial values of R and values after various accelerated weathering cycles are given.

The question of how good a coating needs to be has been addressed by Klopfer, who has established a minimum acceptance criterion which appears to be widely accepted in Europe. The criterion states that R should exceed 50 m if a coating is to exhibit good anti-carbonation properties. (Thus, for a coating applied at a dry film thickness of 150×10^{-6} m, μ_{CO_2} should exceed 340 000.) R for a 25 mm thickness of a normal ordinary Portland cement concrete is about 5 m.

The variation in performance of systems of the same generic type is apparent in Table 5.2 by examination of the R values for the styrene acrylate systems. The pigment volume concentration (PVC) is critical to the performance. Useful information for the specifier must therefore be specific to commercial product brands and the technical literature of many coatings does now include values for μ_{CO_2} and R.

Anti-carbonation coatings, while having a low permeability to carbon dioxide, should at the same time allow passage of water vapour through the film to avoid build-up of vapour pressure behind the coating. This requirement is assisted by the fact that the molecular size of carbon dioxide is larger than that of water. A further acceptance criterion suggested by Klopfer is therefore that the product $\mu_{H_2O}S$ should be less than 4 m where μ_{H_2O} is the diffusion resistance coefficient for water vapour, measured in a similar manner to that for carbon dioxide. However, the factors determining whether a coating will blister are many and complex. Well-bonded simple protective coatings of known low water vapour permeability have given good service without blistering from vapour pressure effects.

An example of a successful use of an anti-carbonation coating was the treatment of a reinforced concrete factory building in 1952 (Lindley, 1987). The building was constructed in the 1930s. In 1952 spalling caused by reinforcement corrosion was repaired with a sand–cement mortar and the building was coated with two coats of an oleoresinous textured coating. A further coat was applied in 1963 to freshen up the surface. Core samples were taken in 1983 from the repaired areas, from a coated unrepaired area and from the interior of the building which had been left uncoated (Fig. 5.16). Carbonation depths measured on the cores were as follows: unrepaired coated area, about 16 mm; repaired coated area, zero; internal uncoated area, about 50 mm. These results indicate that the coating applied in 1952 had arrested carbonation in the original concrete and prevented carbonation in the repair.

There is evidence that hydrophobic pore-lining and pore-blocking materials are not effective in controlling carbonation (Robinson, 1987b). Indeed, the ability of hydrophobic materials to maintain pores comparatively free of moisture may result in optimum conditions for carbonation. Some systems using silane-based treatments also incorporate an anti-carbonation top coating, such as an acrylic resin-based paint, to control carbonation.

5.10 Surface treatments to control moisture or chloride ingress

There is insufficient evidence to date to determine whether, in a wet environment, surface treatments are able to maintain concrete in a sufficiently dry state such that corrosion of reinforcement in concrete already contaminated by chlorides is considerably reduced. Similarly, for concretes which potentially may suffer from ASR, further investigation is needed into the ability of treatments to control the moisture content of the concrete sufficiently that the destructive expansion phase is restricted. Silane type treatments can in certain circumstances reduce the rate of development of ASR damage, and have been used to apparently good effect in Iceland for this purpose. There is a role for treatments to prevent salt ingress to ASR-affected structures which otherwise would exacerbate the problem (Wood, 1985).

It is not necessarily desirable, as mentioned in section 5.9, that surface treatments should be impermeable to water vapour. Concrete should be allowed to 'breathe'. Surface treatments, however, can restrict the ingress of water and therefore the penetration and movement of chloride ions, which would otherwise be drawn into solution in concrete by capillary action or diffuse through water-filled pores.

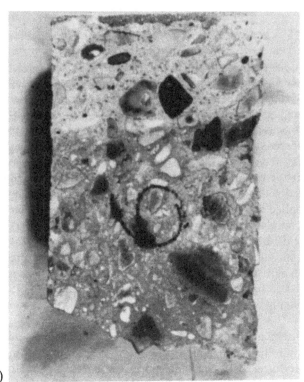

(a)

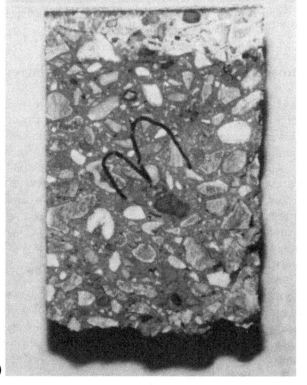

(b)

(c)

Fig. 5.16 Cores extracted in 1983 from a 1930s reinforced concrete factory building: (a) uncoated concrete from interior surface; (b) concrete coated in 1952; (c) core from an area repaired with mortar and coated in 1952. The cores have been sprayed with phenolphthalein solution; lighter areas indicate depth of carbonation. (Courtesy of W.J. Leigh & Co.)

Treatments used for this purpose are drawn from all classes: coatings and sealers; pore-liners; pore-blockers; renderings. A variety of test methods have been used in the laboratory to assess their effectiveness in preventing water or chloride ingress: high pressure permeability tests, weight gains by absorption during immersion, initial surface absorption tests (ISAT) , moisture penetration under shallow pondings, chloride diffusion measurements through thin cement paste or mortar sections as well as monitoring of the onset and rate of corrosion of treated reinforced concrete specimens. Full details of most of these test procedures can be found in a Concrete Society Technical Report

(1988). Results from such tests are frequently provided by suppliers in their trade literature.

Also quoted in some trade literature is a report from the Transportation Research Board in the USA on the comparative performance of 21 surface treatments in preventing chloride ingress into concrete (National Cooperative Highway Research Program, 1981). Tests were carried out on 100 mm concrete

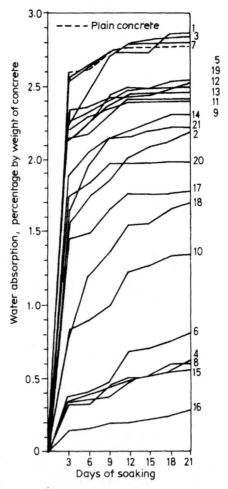

Fig 5.17 Weight gains of concrete with various surface treatments after immersion in 15 wt% sodium chloride solution (NCHRP, 1981): 1, siloxane; 2, boiled linseed oil; 3, siliconate; 4, 14, two urethanes; 5, chlorinated rubber; 6, silane; 7, 11, two materials based on butadiene; 8, 10, 13, three methacrylates; 9, silicate; 12, based on isobutylene and aluminium stearate; 15, 16, 18, 19, 21, five epoxies with different amounts of solids; 17, 20, two epoxies containing polysulphides.

cubes by immersing the specimens after treatment in a 15% by weight sodium chloride solution and recording the weight gain over a period of 21 days, after which the chloride content of the specimens was determined. Treatments tested included coatings and sealers (epoxies, methacrylates, urethane, linseed oil, chlorinated rubber and butadiene-based materials), hydrophobic pore-lining treatments (siliconate, silane, siloxane, stearate) and a pore-blocking silicate treatment. Figure 5.17 indicates the relative weight gains. Chloride ion contents indicated approximately the same relative performance. Once again it is important to note the variability in performance of materials of the same generic title, e.g. epoxies 16 and 19, methacrylates 8 and 13. This is emphasized in Fig. 5.18 from work on silane-treated concrete by Robinson (1987a) where the effects of four types of silane treatment on the time to the onset of steady state diffusion and on the steady state diffusion rate are shown. The comparatively poor performance of treatment type I is probably due to the degradation of the treatment by alkali attack. Similar variability in performance between different emulsion paints has been reported by Whiteley (1977) using the ISAT method.

Thus, as with coatings, useful data on performance for engineers contemplating surface treatments to resist water or chloride ingress must be

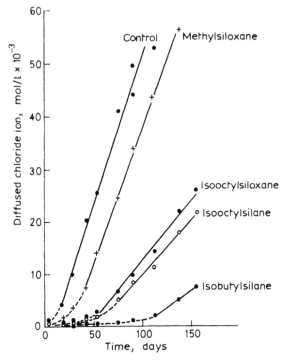

Fig. 5.18 Diffusion of chloride through silane–siloxane-treated concrete (Robinson, 1987) (full line is steady state diffusion).

product specific and, to this end, as the CIRIA Report (1987) recommends, certified test methods must be developed so that there is a common basis on which an informed choice can be made. The methacrylate coating and silane treatment which had performed well in Fig. 5.17 were part of a further investigation in the USA on various corrosion protection systems in which reinforced slabs were subjected to cyclic chloride ponding (Pfiefer *et al.*, 1987). The silane-treated concrete was again very effective in preventing chloride ingress, but in this instance the methacylate coating was ineffective. An examination of the coated surface after 48 weeks of cyclic testing showed evidence of disintegration of the coating. Field evaluation of silane treatments to bridge decks in the USA has indicated significant reductions in moisture and chloride penetration into the concrete. Also in the USA, the technique of overlaying deteriorating chloride-affected bridge decks with high quality or latex-modifed concrete has been widely used (Hay and Virmani, 1985). A test programme investigating the effects of a number of overlays or treatment types (including a silane material) applied to chloride-contaminated slabs indicated that, while some systems reduced corrosion activity, none stopped rebar corrosion in the underlying salty concrete.

The pore-blocking treatments which are applied as cement-based slurries to the concrete surface, and which also act as renders (if the render layer does not flake off), have a practical history of effective action in preventing permeation of water under pressure through basement walls, for example. However, if the rendered layer is removed so that only the pore-blocking that has occurred in the substrate is assessed, such treatments are found to be much less effective in controlling the capillary absorption of water and water-borne chlorides. Silicate and silicofluoride liquid treatments are similarly ineffective in this respect.

The hydrophobic pore-lining treatments, by controlling the moisture content and chloride ingress in concrete, can be effective in improving the frost and freeze–thaw resistance of concrete. They have been used for this purpose in West Germany in particular, with apparent success. Figure 5.19 indicates the difference, recorded in 1985, between treated and non-treated concrete areas on the hard shoulder of a bridge in West Germany which had been treated with an isobutyltrimethoxysilane in 1972. Suppliers' literature reports cyclic freezing–thawing laboratory tests in which the weight loss of concrete slabs treated with a silane-type product is considerably reduced compared with that of untreated slabs (Wacker Chemie, 1985). Field and laboratory tests undertaken in North America to determine the effect of concrete-curing and surface-sealing compounds (including linseed oil, epoxy, silicone and chlorinated-rubber-based materials) on the resistance of air-entrained and non-air-entrained concrete to freezing and thawing in the presence of de-icing salts concluded that the performance of sealing compounds was not impressive. The most important single factor in the attainment of durable concrete was that the concrete was properly air entrained (Ryell and Chojnacki, 1970).

Fig. 5.19 Freeze–thaw effects on silane-treated concrete (right) and untreated concrete (left) on the hard shoulder of a bridge structure in West Germany (courtesy of Dynamit-Nobel)

5.11 Concluding remarks

The potential of surface treatments to concrete to give added protection against deterioration arising from reinforcement corrosion or freeze–thaw damage is now widely recognized. While recent interest has principally been to enhance the durability of existing structures in danger of serious deterioration, surface treatments are being increasingly specified for new structures, particularly in what are recognized to be critically vulnerable areas. This implies, of course, continuing maintenance on a regular basis. Laboratory test results are available which demonstrate the benefits which can be obtained in resisting carbon-dioxide, water and chloride ingress. The specifying engineer, however, needs product-specific data since the performance of materials within a generic grouping can be variable. Suppliers are providing supporting test data on which comparative judgements can be made between materials (where test methods adopted are the same), but evidence of proven and successful performance in service over a period of years for the currently available formulations is ideally required but less readily obtained. Excellent laboratory results and a successful track record for a treatment will mean nothing, however, if the treatment is not applied correctly to the concrete surface of the job in question, to specification and under supervision. Nor should surface treatments be regarded as substitutes or alternatives to the achievement of high quality, well-compacted and cured concrete. A cover zone to the thickness and quality specified in codes of practice will be adequate in most situations, but the engineer may judge that the added protection afforded by a surface treatment is sometimes necessary.

References

ACI (American Concrete Institute) (1966) Guide for the protection of concrete against chemical attack by means of coatings and other corrosion-resistant materials. *J. Am. Conc. Inst.*, **63** (12), 1305–91 (Committee 515).

ACI (1985) *Manual of Concrete Practice Part 5*, ACI 515. IR-79, 515.IR 1–515.IR 41, American Concrete Institute.

Anon. (1978) Waterproofing concrete. *Civil Engineering (inc. Public Works Review)*, May, pp. 57–9, 71.

Berkeley, K.G.C. and Pathmanaban, S.(1989) *Electrochemical Corrosion Control in Concrete*, Butterworths.

BRE (Building Research Establishment) (1977) Painting Walls: Part 1, Choice of paint, Part 2, Failures and remedies. *Digests 197 and 198*, BRE, Garston.

BSI (British Standards Institution) (1982) BS 6150: *Code of Practice for Painting of Buildings*, BSI, London.

CIRIA (Construction Industry Research and Information Association) (1987) *Technical Note 130*, CIRIA, London.

Concrete Society (1988) Permeability Testing of Site Concrete — a Review of Methods and Experience, Report of a Working Party, *Technical Report 31*, Concrete Society.

Concrete Society (1989) Cathodic Protection of Reinforced Concrete, Report of a Working Party, *Technical Report 36*, Concrete Society and Corrosion Engineering Association.

Department of Transport (1986) *Materials for the Repair of Concrete Highway Structures*, Departmental Standard BD 27/86, Department of Transport, London.

Desor, U. (1985) Protection and restoration of concrete with emulsion paints. Experiments on carbon dioxide and sulphur dioxide permeability. *Symposium on Coatings for concrete*, Paper 3, Paint Research Association, pp. 14–18.

Edmeades, R.M. and Hewlett, P.C. (1985) Waterproofing admixtures for Concrete. *Concrete*, **19** (8), 31–2.

Engelfried, R. (1985) Preventive protection by low-permeability coatings, *Permeability of Concrete and its Control*, Concrete Society, London, pp. 107–17.

Fletcher, T. (1982) *New Civil Engineer*, October, pp. 28–9.

Hankins, P.J. (1985) *Proceedings of the 1st International Conference on Deterioration and Repair of Reinforced Concrete in the Arabian Gulf*, Bahrain Society of Engineers, pp. 273–85.

Hay, R.E. and Virmani, Y.P. (1985) North American experience in concrete bridge deterioration and maintenance, *Concrete Bridges, Investigation, Maintenance and Repair*, Concrete Society, London, pp. 14.

Heuze, B. (1980) Cathodic protection on concrete offshore platforms, *Materials Performance*, **19**, 24–33.

Jones, G.H. (1983) Surface coatings for concrete: a comparison of effectiveness in preventing carbonation, *BRE Note N 147/83*, BRE, Garston.

Kish, G. (1981) Control of stray current on reinforced concrete, *Corrosion '81*, Paper 51, National Association of Corrosion Engineers, Houston, TX.

Klopfer, H. (1978) Carbonation of exposed concrete and its remedies. *Bautenschutz Bausaneriung*, **1** (3), 86–96.

Leach, J.S.L. *et al.* (1987) The cost in primary materials and energy of metallic corrosion in the EC, *Proceedings of the European Federation of Corrosion*, April, Federal Republic of Germany.

Lindley, R. (1987) Painting and protecting concrete, *Symposium on Protection of Concrete and Reinforcement*, Paint Research Association/Cement and Concrete Association, Slough.

McCurrich, L. and Whitaker, G. (1985) Reduction in rates of carbonation and chloride ingress by surface impregnation. *Proceedings of the 2nd International Conference on Structural Faults and Repair*, Engineering Technics Press, Edinburgh, pp. 133–9.

Moller, L. (1985) Research on protection of steel reinforced concrete in Scandinavia. *Symposium on Coatings for Concrete*, Paper 2, Paint Research Association, pp. 9–13.

National Association of Corrosion Engineers (NACE) (1985) Cathodic protection of reinforced bridge decks, *Proceedings of the Bridge Deck Seminar*, National Association of Corrosion Engineers, Houston, TX.

National Cooperative Highway Research Program(NCHRP) (1981) *Report 244, Concrete Sealers for the Protection of Bridge Structures*, Transportation Research Board, Washington, DC.

Pfeifer, D.W., Landgren, J.R. and Zoob, A. (1987) *Protective Systems for New Prestressed and Substructure Concrete*, Report No FHWA/RD-86/193, US Department of Transportation, Federal Highway Administration, Washington, DC.

Robinson, H.L. (1986) Evaluation of coatings as carbonation barriers. *Proceedings of the 2nd International Colloquium on Materials Science and Restoration*, Technische Akademie Esslingen, West Germany, pp. 641–50.

Robinson, H.L. (1987a) An evaluation of 'silane' treated concrete. *J. Oil Colour Chemists Assoc.*, **70**, 163–72.

Robinson, H.L. (1987b) Durability of anti-carbonation coatings. *J Oil Colour Chemists Assoc.*, **70**, 193–8.

Roth, M. (1986) Siliconates — silicone resins — silanes — siloxanes, *Seminar on Concrete and Masonry Protection*, Thames Valley Section, Oil and Colour Chemists Association.

Ryell, J. and Chojnacki, B.(1970) Laboratory and field tests on concrete sealing compounds. *Highway Research Record 328*, pp. 61–76, Highway Research Board, Washington DC.

Shaw, J.D.N. (1980) Special finishes for concrete floors. *Advances in Concrete Slab Technology* (eds. Dhir, R.K. and Munday, J.G.L.) Pergamon Press, Oxford, pp. 505–15.

Sherwood, A.F. and Haines M.J. (1975) The carbon dioxide permeability of masonry paints. Paint Research Association. *Technical Report*. TR/1/75, PRA, Teddington, Middlesex.

Turner, G.P.A. (1988) *Introduction to Paint Chemistry and Principles of Paint Technology*, 3rd edn, Chapman and Hall, London.

Wacker Chemie GmbH (1985) *Technical Literature Test Report 234/85/LMC*.

Washa, G.W. (1933) Efficiency of surface treatments on the permeability of concrete. *J. Am. Conc. Inst.*, **5**, 1–8.

Whiteley, P. (1977) The performance of organic coatings on elements of external walls, *Proc. RILEM/ASTM/CIB Symposium, Finland*, pp. 410–23.

Wood, J.G.M. (1985) Engineering assessment of structures with alkali-silica reaction *Proceedings of Conference on Alkali–Silica Reaction*, Concrete Society, London.

Conclusions 6

G.C. Mays
(Royal Military College of Science, Cranfield)

6.1 The future

The concrete construction industry in the UK having lived through the high alumina cement and calcium chlorides scares of the 1960s and 1970s was then faced with the problem of alkali–silica reaction in the 1980s. The first two phenomena have been dealt with by prohibition of the use of certain materials in structural concrete and with the latter at least the mechanisms involved are now reasonably well understood even if there is no cure available at present. Other internal processes which can lead to deterioration, such as sulphate attack and thermal crack generation, are also basically understood and in each case a number of design or construction options are available to reduce the chance of problems in service. A full explanation of the mechanisms involved in each case has been provided in Chapter 2.

As long as the materials currently being used in the manufacture of concrete or any new materials subsequently introduced do not produce another scare in 20 or so years' time, the main causes of deterioration in the future are likely to be as a result of the environment in which the structure is built. Here the concern is the penetration of salts, oxygen, moisture and carbon dioxide. With regard to the last three agents, there is no excuse today for ignoring the basic design and construction prerequisites for avoiding carbonation and subsequent reinforcement corrosion. Indeed there has been no excuse for many years. The 1957 version of British Standard CP114 stated that

in order to ensure adequate protection to the reinforcement with the cover specified in Clause 307, it is essential that it should be dense, impermeable and of a quality for the conditions of exposure involved. The greater the severity of the exposure, the higher the quality of concrete required, and thus although a weaker concrete may be adequate from structural considerations this weaker mix may not be suitable from the durability view-point. (BSI, 1957)

Further, in their handbook on the Code, Scott *et al.* (1965) state that concrete strength is not the primary feature in providing protection to the steel, the more fundamental properties being cement content, water–cement ratio, workability and degree of compaction. With the benefit of hindsight, it is quite incredible that it was not until 1985 that BS 8110 (1985) formally incorporated the former basic principles. Perhaps what was crucially omitted earlier was the importance of curing and this has become particularly important today when using partial replacement of ordinary Portland cement (OPC) with secondary cementitious materials such as ground-granulated blastfurnace slag (ggbfs) or pulverized fuel ash (pfa).

With regard to the penetration of salts, particularly chlorides, in solution it is perhaps necessary to consider additional techniques at the time of construction to make the reinforcement more resistant to corrosion. These might include the use of epoxy-coated reinforcement, the alternative of stainless steel being very expensive (Tilly, 1988). For protecting structures which have been exposed to chlorides but in which the levels are not yet high enough to cause active corrosion, the application of protective coatings as discussed in Chapter 5 may be the best option. Where chlorides are present and corrosion is active Tilly suggests that cathodic protection, also described in Chapter 5, is the most feasible option for strategically important structures. In both areas, however, there is still a need for a considerable amount of development work backed up by field trials.

There will remain a significant number of structures which, from time to time, will need repair in the form of local patching or crack sealing. The materials and the techniques available are thoroughly reviewed in Chapter 4. For larger volume repairs the use of the superfluid microconcretes is now finding considerable favour and these, appropriately formulated, are less likely to give problems due to property mismatch than some of the polymer or polymer-modified systems. Nevertheless, the load state of the structure during and subsequent to the repair process, together with the use (or otherwise) and the efficiency of propping systems are problems which will tax the minds of those involved in the design of major repair schemes in the future.

6.2 Maintenance policies

Concrete, like other construction materials, is not maintenance free and the future will require the development of realistically costed maintenance policies allied to appropriate repair strategies. The case studies in Part Two of this book all show that the owners of our building and civil engineering infrastructure have taken this fact on board very seriously. Maintenance policies applied to existing structures generally involve the decision to implement one of three basic repair philosophies (Brown *et al.*, 1985).

1. Do nothing other than to incorporate measures to protect public safety and accept significant reduction in structure life (Fig. 6.1).

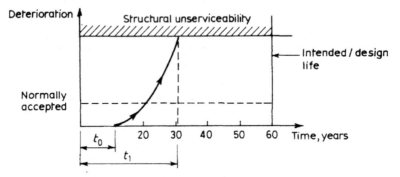

Fig. 6.1 'Do nothing' repair philosophy (t_0, time to initiation of damage; t_1, actual life).

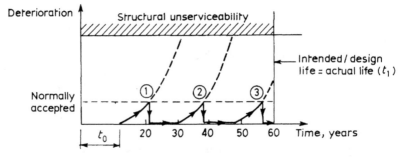

Fig. 6.2 Regular holding repairs.

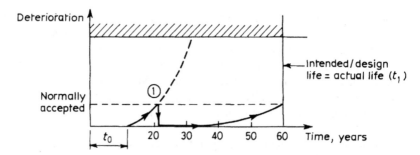

Fig. 6.3 Once-off full refurbishment.

2. Carry out holding repairs to slow down corrosion rate, accepting the need for further repairs at intervals in order to reach the original intended life of the structure (Fig. 6.2).
3. Carry out a once and for all major refurbishment, perhaps using surface treatments and/or cathodic protection, designed to enable the structure to survive for its intended life (Fig. 6.3).

Further options include demolition, sale or the complete replacement of affected members. The choice will be closely allied to the funds available for maintenance and how these are distributed over time.

Whatever the decision, the strategy must be based on a thorough structural investigation, the prime aim of which must be to discover the cause of the distress or deterioration. A number of test and inspection techniques as described in Chapter 3 may be employed. In interpreting the results of the site and laboratory work it is important not to rely on one method on its own but to deduce a consensus from the wide range of evidence often available.

At the risk of repeating what has been said several times before, concrete which has been properly designed and constructed to suit the environment in which it will operate will prove to be a very durable material given that it receives a reasonable level of routine preventative maintenance throughout its design life.

References

Browne, R.D. Gibbs, B. and Robery, P.C. (1985) Repair of concrete structures subject to environmental attack, Supplementary paper presented to the 2nd International Conference on Structural Faults and Repair, London.

BSI (British Standards Institution) (1957) CP 114: *Code of Practice for Reinforced Concrete*, BSI, London.

BSI (1985) BS 8110: *Structural Use of Concrete. Part 1: Code of Practice for Design and Construction*, BSI, London.

Scott, W.L. Glanville, W. and Thomas, F.G. (1965) *Explanatory Handbook on the BS Code of Practice for Reinforced Concrete CP114: 1957*, Concrete Publications, London.

Tilly, G.P. (Feb 1988) Durability of concrete bridges, *Highways and Transportation*, **35** (2), 10–19.

PART TWO
Case Studies

High-rise buildings 7

C. Stroud and P. Gardner *
(London Borough of Newham)

7.1 Introduction

Many of today's high-rise buildings in the UK were built in the boom of the 1960s, some using relatively new technology. The emphasis was often on low capital cost and fast rates of construction, with the significance of future maintenance not being appreciated. Workmanship and site supervision were generally inadequate for both the new technologies used and the speed of construction.

Many high-rise housing blocks are now presenting their owners with substantial maintenance and repair problems, in particular in the area of concrete construction, which has not always proved to be the durable maintenance-free material that it was assumed to be.

The London Borough of Newham has approximately 115 high-rise housing blocks of eight storeys and above. These contain over 7000 units, nearly one-third of the Council's housing stock. The majority of this stock will have to continue to provide at least adequate housing for some way into the future.

It is clearly not sufficient to wait for significant defects to become apparent and only then respond to these problems. Planned maintenance and property asset management could limit deterioration, uplift standards of existing buildings and result in an increased return on investment. However, the level of available financial resources is often well below that required for adequate maintenance, let alone repair or refurbishment schemes.

The prerequisite to any repair or refurbishment programme is a knowledge of the buildings concerned derived from appropriate investigations. A major part of these investigations has been Newham's five year rolling programme of tall block inspection.

* This chapter is based on a paper presented by the authors at a conference on concrete repair in London on 16 March 1988 and is published with the permission of Palladian Publications Ltd. Prices quoted are those current at that date.

7.2 Five year tall block inspection programme

7.2.1 Evolution of the inspection programme

Immediately after the partial collapse of the Ronan Point building following an accidental gas explosion in 1968 (Chapter 1), all high-rise housing blocks in the Borough were checked for adequacy of strength in relation to wind and gas explosion loading. In these surveys only two buildings of similar construction were found to be deficient. However, whilst these were being strengthened, the Council became aware of other structural problems with its high-rise stock. These included failure of slip bricks and mosaic finishes, inadequately fixed cladding panels and spalled concrete.

The decision was taken that all blocks over eight storeys would be structurally inspected on a five year rolling programme. These initial surveys were carried out using binoculars, supplemented by inspectors on bosun's chairs. A small amount of work in which men gained access from cradles suspended from the top of the building was necessary. Many defects in materials and workmanship were found resulting in an expenditure of £9.63 million on remedial works between 1971 and 1985, including the demolition of two 16 storey blocks.

In the early 1980s it was apparent that the inspection techniques were not adequate. Additional work was often identified once the building had been scaffolded and this led to considerable extra expenditure on contracts after they had been let. The Engineering Department was criticized for this and the resulting reduced budget for other schemes in the Authority's Capital Programme.

In conjunction with the Housing Department it was agreed that the surveys needed to give more precise information on the nature and extent of defects and on the recommended remedial works.

7.2.2 The client brief for the tall block inspection

No survey can cover every eventuality; many of the construction details are buried and can only be exposed at great expense and with much aggravation to the building's occupants. There has to be a compromise based on the level of information reliability required, the resources which the client is willing to spend and the disruption which is acceptable.

After discussion with the Housing Department and the Housing Committee, the following basic brief was agreed:

1. The survey should be limited to the external envelope and communal areas. Access to individual flats would not be allowed.
2. The survey should identify all defects affecting public safety.

3. Structural details should only be reviewed either if there were signs of visible distress or if similar details had already caused problems elsewhere in the industry.
4. The survey should enable each building to be prioritized for future inspection, maintenance, repair and remedial work. It should identify all the major technical problems that would need to be addressed in an external envelope refurbishment.

7.2.3 Methodology

Staffing

The Councillors wanted as much as possible of the survey, the design of the remedial works and the subsequent site supervision to be carried out by in-house staff. However, when the inspection programme commenced in 1982, insufficient internal resources were available and so the programme was split between two consultants, with a decision to recruit and train staff to take over the role in future years. It was felt particularly important that during training the in-house staff should gain practical experience in viewing the defects being located at first hand.

External access

The external deterioration of the buildings was obviously going to be a continuous problem. It was therefore decided that, as the first inspection of the programme would be the basis for future cycles, each building should be inspected at 'touch' distance where possible. To achieve this the majority of the work was carried out with men gaining access using cradles provided under an annual contract for the provision of access equipment and associated labour. Specialist contractors using abseiling techniques have been used where access has been particularly difficult or where only monitoring of known defects has been required.

The survey teams include two inspectors who carry out the physical surveying and testing. They are supported by in-house structural engineers who visit the site regularly and prepare the final reports, which include recommendations for remedial works.

Inspection check list

The general brief given to the inspectors with regard to concrete is to record all deterioration on sketch elevations and by photographic record. A covermeter survey is to be carried out on exposed concrete surfaces and recorded on sketch elevations. The depth of carbonation is to be assessed using phenolphthalein indicator (Chapter 3). Samples are to be sent for laboratory testing to identify any chloride problems.

Specific brief

In addition to items covered in the above check list, a specific brief is agreed for each block by the engineer before inspection commences. This covers items unique to the block or block type, the initial number of cradle drops, the elevations to be inspected and any other special considerations. The number and position of cradle drops may be varied following information gained from the initial elevations inspected.

7.2.4 Development of the tall block inspection brief

Many changes have been made to the inspection procedure, the equipment used, the survey techniques and the extent of detail required. A number of critical areas merit consideration.

Tenant knowledge

Since the onset of the programme the Housing Service of Newham has been decentralized. As a result tenants offer much more information on defects in their homes. It became obvious that the original brief of not entering the flats was inadequate. Some tenants have good knowledge of the block's remedial works history and of the development of particular cracks or defects. They were also concerned over the lack of reference to insulation and condensation problems or the adequacy of heating provision.

The basic brief has since been modified to include internal inspection of all top storey flats and any other flats where tenants have indicated possible problems.

Archive information

The extraction of archive information has proved to be far more problematic than originally anticipated. No relevant structural drawings have been located for around 50% of the buildings and, whilst the London Borough of Newham's problems may be individual, the cost effectiveness of archive retrieval should be a matter of serious consideration elsewhere. Even if the drawings can be found, one cannot be sure that they represent what actually exists on site.

There seems to be a sound case for the establishment of 'log books' for individual buildings to assist in future inspections, maintenance and even demolition.

7.2.5 Typical structural defects identified by the inspection programme

There is considerable overlap both of type and cause of the defects so far identified and they can be conveniently grouped under three main headings:

1. concrete and concrete finish deterioration;
2. cladding materials and their workmanship;
3. movements in the structure.

As far as concrete deterioration is concerned, corroding reinforcement causing spalling of the concrete is an extensive problem and the need for concrete repair is often the catalyst for a major remedial contract. There are four main influencing factors: cover, carbonation, chlorides and exposure.

The low cement 'designed' mixes of the 1950s and 1960s were sometimes highly permeable and consequently can be carbonated down to the steel even where concrete cover is quite high. However, the common problem is essentially that of insufficient cover to reinforcement. There are vast areas of concrete with under 10 mm of cover.

Problems caused by chlorides are relatively rare in Newham and are generally limited to replaceable items such as precast walkway columns, window surrounds and cladding units.

A variety of finishes to concrete surfaces have been used and the most common are mosaic, exposed aggregate and cobblestone. Mosaic, owing to moisture expansion, lack of initial adhesion and frost, has been an absolute failure and has necessitated considerable expenditure in either its removal or overcladding. It is felt that the exposed aggregate and cobblestone finishes, whilst currently sound, may pose considerable problems as a result of detachment in the future.

7.3 High-rise repair strategies

There is generally no single remedial solution to the problems outlined above. There are a number of differing strategies open to the Borough, however, which will be considered in turn.

7.3.1 Public safety (emergency) works

This represents the minimum option open to the Council. The Works comprise primarily the removal of danger to the public and occupants. Most commonly this danger manifests itself in the form of falling debris, predominantly from spalling concrete although on occasions from loose or unstable masonry or cladding panels.

One solution is the provision of protective scaffold fans to catch falling debris. However, the Borough rarely uses protective fans for the following reasons:

1. vandalism and misuse of the scaffold;
2. cost of hire if repairs are not carried out within the short term;
3. maintenance of the fans;
4. tenant inconvenience/privacy;
5. tenant security problems.

The preferred solution is to hammer test all suspect concrete surfaces and remove any loose material. The removal of loose concrete by hammer testing is only a temporary solution with the sole purpose of removing immediate danger to the public. The works are short lived and unless full repairs are carried out subsequent emergency works are likely to be required on a regular and increasing basis as the deterioration will accelerate as a result of the physical removal of concrete cover.

The access to carry out the works is normally by cradles although abseiling has been used on areas of difficult access.

The minimum option has until recently been extensively used in the Borough with some £100 000 per annum being spent.

7.3.2 Holding repairs

The Borough, conscious of its high expenditure on emergency works and not having the resources to carry out an extensive programme of full refurbishment contracts, has embarked on a programme of holding repairs.

This programme has five main objectives:

1. to reduce the need for emergency works to tower blocks;
2. to incorporate urgent repairs on a planned basis;
3. to delay the need for further works for up to five years (when it is hoped a full refurbishment contract can be carried out);
4. to reduce the rate of deterioration of the structure;
5. to allow structural works to be carried out in isolation without the need for a full refurbishment contract.

The works to concrete surfaces comprise the hammer testing of all exposed concrete and the subsequent grit blasting of all spalled areas. The exposed steel reinforcement is then treated with a suitable proprietary primer and the adjacent concrete areas are treated with a carbonation-resisting sealant. Particular emphasis is placed on vertical surfaces such as floor slab edges which tend to be more affected than soffits. Access is normally by cradles although limited scaffolding may also be employed.

The Borough currently spends in the region of £500 000 per annum on holding repairs and carries out works to some 15–20 blocks each year. The annual Holding Repairs Contract essentially consists of concrete repairs.

The majority of roof and brickwork repairs are permanent and include the provision of new roof membranes, supplementary wall ties and crack stitching.

The need for works to individual blocks is identified from the Tall Block Inspection Programme and then prioritized against the funding available to form a programme for the financial year. Contract documents are then produced which give broad details of the works required and individual Bills of Quantities for each block. Because of the nature of the works, the Bills of Quantities are provisional and are essentially schedules of rates with guide quantities for pricing. It is not until the hammer testing and grit blasting is completed that the full extent of concrete deterioration can be quantified.

The quantities of concrete repairs to a block are based upon the total area of exposed concrete and an engineering judgement of the proportion likely to be degraded, this judgement being based on the tower block inspection reports (which give details of concrete cover, carbonation depths and incidences of spalling) and experience gained from previous projects. In Newham this judgement is further complicated by the fact that virtually all exposed concrete surfaces have been decorated with a simple masonry paint.

Rates are contained within the Bill of Quantities for both holding and full concrete repairs. Upon the completion of hammer testing and grit blasting the total cost to complete the block can be obtained. If funds permit or if the works are less extensive than originally billed the opportunity can be taken to carry out full repairs or to provide preventative maintenance in the form of more extensive over-coating with carbonation-resisting paints.

Very close financial control is kept by carrying out week by week cost analysis. By this cost control it has been found possible to maximize the effectiveness of the contract but still remain within budget.

7.3.3 Full refurbishment

The decision to carry out a full refurbishment scheme to a block is usually led by the need to carry out substantial structural repairs. In a full refurbishment contract, however, all aspects of the block are considered and generally the works include many of the following:

1. full repair to the external envelope including roof works;
2. window replacement (upgrade or renewal);
3. services (upgrade or renewal);
4. lifts (upgrade or renewal);
5. internal works to flats (bathrooms, kitchens etc.);
6. works to communal areas (redecoration, environmental works);
7. door entry systems (or other security works);
8. asbestos removal;
9. heating replacement.

The cost is usually high: up to £30 000 a unit or £1 million for a typical eight storey block of 32 flats. The Council currently carries out two to three full refurbishment contracts per year to tall blocks.

The works normally require the complete scaffolding of the block. Because of the high costs of access scaffold and site set-up, the opportunity is usually taken to replace or repair any aspects which have a limited life, e.g. to replace timber windows with UPVC units. Although this makes economic sense in the longer term, it does utilize funds which could have been allocated to other buildings in need of repair.

7.3.4 Demolition

Certain blocks, because of major structural faults and/or the need for extensive works to services, lifts, flats etc., may not be economic to refurbish. Social aspects may also be a factor leading to the decision that demolition is the most economic answer.

7.3.5 Sale

Blocks which are not economic to maintain in the public sector may be a more viable proposition in the private sector. In these cases, consideration may be given to their disposal by sale. Alternatively, blocks with severe social problems may be sold to specialist housing associations.

Increasing property values compared with refurbishment costs resulted in many of the Borough's blocks becoming economic to refurbish. However, the Council does not necessarily have the necessary resources and, in these cases, sale to private developers can lead to the Council gaining substantial capital receipts or units of accommodation in barter deals.

7.4 Repair and maintenance strategies

7.4.1 Preventative maintenance

The high cost of concrete repair should logically lead to the decision to provide specialist carbonation-resisting paint systems for all exposed concrete surfaces as preventative maintenance. Unfortunately, because of the Council's extensive list of urgent repairs this cannot yet be prioritized although a limited amount is currently being undertaken as part of the holding repair programme.

There is a strong argument, however, that in the long term it may be more effective to carry out preventative maintenance on those blocks currently in sound condition at the expense of, say, other full refurbishment schemes.

However, these decisions are complex and involve a balance between the various strategies in terms of distributing the expenditure available. To make these decisions a large number of criteria have to be considered.

7.4.2 Criteria for strategies

To decide on the most appropriate course of action in the repair and maintenance of high-rise housing, a number of criteria have to be considered, including the following.

Social aspects/housing management	1. Tenant profile 2. Size of units 3. Location 4. Block height 5. Amenities
Structural works needed	1. Urgency of works 2. Extent of works 3. Cost of works 4. Access requirements 5. Requirements for alternative accommodation
Other works needed	1. Lifts 2. Windows 3. Rubbish chutes 4. Services 5. Security 6. Roofing 7. Heating 8. Asbestos 9. Insulation
Funding	1. Funds available 2. Timescales 3. Priorities
Political environment	1. Tenant pressure 2. Council Members' policy requirements 3. Central Government policy

From these criteria, for example, decisions may be made that a particular building has no long-term future in the Council's housing stock owing to social aspects. Consequently, only public safety works would be carried out and not full refurbishment. Alternatively, a block with extensive structural problems may be the subject of holding repairs to allow sufficient time for movement of tenants to alternative accommodation to facilitate a future refurbishment contract or sale of the block.

The structural survey detailed earlier will not provide information on all the above criteria. Other initiatives have to be taken to allow a fully rational decision to be made on any building's future. To enable this to occur, the Borough recently set up a multidisciplinary team which undertook a structural survey, a social survey and a flat condition survey of every tower block in the Borough. From this information decisions can be made on the future of each tower block and its maintenance needs both in the short and longer term. This then allowed the Borough's holding repair and full refurbishment programmes to be formulated.

Two case studies follow to illustrate how some of the above aspects have led to the Council's current solutions to the problems of high-rise housing repair and maintenance.

7.5 Case studies

7.5.1 Thomas North Terrace

Thomas North Terrace is an 11 storey building consisting of 80 flats (Fig. 7.1). The building is ideally situated in terms of transport, community facilities and shops. The tenancy profile is mainly elderly singles and couples and the block works reasonably well socially.

The building comprises *in situ* reinforced concrete floors and party walls clad generally in brickwork. The west elevation has exposed concrete string beams at floor and sill levels with interconnecting columns. The east elevation comprises cantilever concrete public access balconies with precast concrete posts and handrails forming a non-structural balustrade system. The north and south elevations are faced in brickwork with slip bricks covering the floor slab edges.

The structural survey carried out in 1984 reported that the building had extensive and severe concrete deterioration. The string beams and balcony edges were found to have very low cover. In a number of cases the links were actually flush with the outside of the concrete surfaces. At the time of construction these links were red leaded and then painted as part of the original painting scheme for the building. Carbonation depths were found to be in excess of 30 mm in many locations, indicating poorly compacted and porous concrete.

The precast concrete handrails and posts were found to have deteriorated considerably and to contain chlorides. The brickwork was found to be generally

Fig. 7.1 Thomas North Terrace.

deficient in wall ties and the roof, services, lifts and interior of the flats were also in need of upgrading.

The cost to refurbish the block fully was estimated at £1.5 million (1985) and, in view of its location and social condition, Council Members agreed to its refurbishment. Since the refurbishment contract could not commence until 1986, emergency works were carried out from cradles to remove all loose and spalling concrete. The exposed reinforcement and surrounding concrete was then sealed with a suitable primer to reduce further deterioration.

The structural works mainly comprise concrete and brickwork repairs, the block being fully scaffolded for access. The carbonation and covermeter surveys confirmed that for this building all exposed concrete surfaces including balcony soffits required grit blasting, repair with a specialist concrete repair system and overcoating with a carbonation-resisting paint.

There were exceptions to this to reduce costs, however.

1. The concrete balcony handrails and posts were removed and replaced in steel.
2. Where concrete repairs were very extensive in terms of depth and area they were first built up using a small aggregate concrete mix incorporating a styrene-butadiene rubber (SBR) additive.

Fig. 7.2 Herbert Willig House

3. In areas of very low cover, additional cover was added by the utilization of expanded metal and the casting of a small aggregate concrete mix incorporating an SBR additive. Overcoating with a carbonation-resisting paint then completed this repair.

7.5.2 Herbert Willig House

Herbert Willig House is an 11 storey block comprising a reinforced concrete frame and floors with a combination of precast concrete and brickwork cladding (Fig. 7.2). The building dates from 1963 and at the time of inspection had a piped gas supply.

The initial binocular survey indicated that there was corrosion damage to the external concrete columns, balconies and cladding. Close inspection from cradles included hammer testing of the zones of cracked concrete, which caused the detachment of a large portion of a cladding panel. The damaged zones were widespread and some were directly above public access areas. Emergency measures were therefore taken to removal all loose material and provide safe access to the building.

An archive search together with tests for chloride content showed that the cladding defects were attributable to inadequate reinforcement in vulnerable sections and to reinforcement corrosion in carbonated concrete containing 0.4%–1% chloride ions by weight of cement. The lack of reinforcement could not be remedied and the extent of corrosion was considered beyond economic repair. It was clear that for long-term use of the building the cladding would need to be replaced.

Fig. 7.3 Failure of cladding units at Herbert Willig House

The failure of the cladding units (Fig. 7.3) exposed details of the structural frame which, on further investigation, was found to be an unusual combination of precast and *in situ* components and not a traditional *in situ* reinforced concrete frame as originally envisaged. Further structural appraisal of the building concluded that the end bays of the building above third floor level would not accept an equivalent static pressure of $17 \, kN/m^2$ without disproportionate collapse. Immediate action was therefore taken to cut off the piped gas supply and provide electric cookers to all tenants.

In the light of these events, the Council decided to move the tenants to alternative accommodation over a one year period, with three-monthly inspections of the cladding panels to remove any loose material. As far as the future of the block is concerned, the external cladding could be replaced and this would possibly be done if it were the only problem. However, the relatively high cost of structural strenghthening makes the block unsuitable for retention as part of the Council's long-term housing stock.

7.6 Concluding remarks

Concrete deterioration is posing many building owners with a severe problem in terms of the repair and maintenance of their stock. Lack of understanding of the need for good workmanship, supervision and detailing in the construction of concrete structures has led to the need for early repair. The deterioration of reinforced concrete is time dependent and it is therefore likely that it will become an increasing problem for many years to come.

For most building owners and in particular local authorities the problem has to be tackled with very restricted budgets. Consequently owners must ensure that such funds are spent in the most effective way. In order to achieve this the following issues must be addressed when undertaking the repair of high-rise structures.

1. As much information about the structure and its function as is economically viable should be collected. It is essential that a comprehensive survey is carried out to identify all the building's defects.
2. Both short- and long-term strategies for the future of the block should be developed. These must be reviewed periodically as they may change in line with available resources and other influencing factors.
3. Where a chosen strategy involves some degree of concrete repair, appropriate materials which will achieve the required objectives should be selected and specified. Often several different proprietary materials are acceptable and it may then be better to give the contractor the option to choose which product he wishes to use, provided that it meets with the engineer's approval.

4. Appropriate and acceptable access methods for carrying out the works should be selected and specified. Again the final choice may be left with the contractor, who should include details in his method statement for carrying out the works.
5. Good workmanship is probably the most important factor in concrete repair. It is therefore essential that the works are properly supervised and not just left to the specialist contractor.
6. When carrying out construction repair the total problem should be addressed, and not just the immediate or most obvious one. For example, spalling concrete is often the most noticeable defect but it may not be the only problem.

Highway bridges   8

K.J. Boam *

(G. Maunsell & Partners)

8.1 Introduction

Both reinforced concrete and prestressed concrete have been used extensively in the construction of highway structures. Such structures are expected to provide good long-term durability and for many years concrete structures were promoted as being maintenance free.

While good durability can be achieved by careful specification, by attention to detail at the design stage and by ensuring that structures are constructed in the way the designer intended, it is doubtful that any concrete structure whether it be reinforced or prestressed can be considered completely maintenance free. At the very least concrete structures require regular inspection and it has to be expected that even the best structures will require some maintenance during the course of their life.

Corrosion of reinforcement or prestressing represents by far the greatest threat to bridges in Britain and it is becoming apparent that there are significant numbers of bridges and other highway structures that are in need of repair on this account. These repairs are going to cost many millions of pounds.

In this chapter the scale of the corrosion problem affecting British bridges and the areas of structures likely to be at risk are discussed. Options for repair are described as well as means of monitoring the performance of repairs and avoiding problems in the future.

8.2 Scale of the problem

Because of the diversity of ownership of, and the responsibility for, bridges and other highway structures in Britain it is doubtful whether anyone knows exactly how many concrete structures there are. Current estimates are that there are over 150 000 bridges of which there are at least 60 000 bridges

---

*This chapter is based on a paper presented by the author at a conference on concrete repair in London on 16 March 1988 and is published with the permission of Palladian Publications Ltd.

where concrete is the main structural medium (Institution of Civil Engineers 1988). The majority of the remainder of the bridge stock comprises masonry and brick structures and there are approximately 25 000 bridges constructed in steel or iron. Concrete has been used in the strengthening of brick and masonry arches and it has also been employed in the sub-structures of many bridges and steel decks. If retaining walls, culverts etc. are included then it is likely that there are over 100 000 highway structures in Britain in which concrete plays a structural role.

Chloride contamination due to the use of road salt is considered likely to be the most serious cause of deterioration of British bridges. Current estimates are that between 1 million and 2 million tonnes of salt are spread on Britain's roads every winter. This means that a 20 m span motorway bridge may receive over 10 tonnes of salt during its 120 year design life! Road users have come to accept de-icing as the norm and the economic and safety implications of not maintaining strategic routes free from ice are such that a change from current policy is unlikely to be contemplated. The use of salt has necessitated major repairs to a number of bridges. In a few cases these repairs have amounted to a virtual replacement. Investigations of many other bridges have shown either that extensive work is needed or that chlorides are present and that corrosion will inevitably result. Still more structures, which have not been subjected to a detailed examination, exhibit the symptoms of chloride-induced corrosion.

A recent study on behalf of the Department of Transport (1989) involving a representative sample of 200 bridges showed that the most common cause of reinforcement corrosion or spalling was inadequate reinforcement cover. Most of the bridges studied contained potentially harmful levels of chloride in the concrete, mainly arising from de-icing salt. However, carbonation was not a major problem on most of the structures. The test results for chloride concentration and half-cell potential were used to classify the bridges into good, fair and poor condition. Sixty were found from testing to be good, 99 fair and 41 poor. Comparative figures based on a visual inspection were 25 good, 114 fair and 61 poor.

Those parts of highway structures likely to be at risk from chloride-induced corrosion are now discussed in turn.

8.2.1 Bridge decks

The widespread use of waterproof membranes on bridge decks in Britain is perceived as providing decks with good protection against the effects of salt. However, the fact that a bridge deck is covered by a membrane and surfacing means that it is difficult to confirm that this is the case.

All waterproofing systems demand a high standard of workmanship to ensure that they are initially watertight. Some membranes undoubtedly

Fig 8.1 Taff Fawr Bridge.

deteriorate with time and all are susceptible to damage during the course of resurfacing operations. (It is only recently that the Department of Transport have made it mandatory for a protective layer to be provided over a waterproof membrane.) Thus, de-icing salt is likely to have found its way through the membranes on some bridges and there will therefore be structures where corrosion is going on unseen beneath the surfacing.

The deck of Taff Fawr Bridge in Wales (Fig 8.1) has recently been replaced because of corrosion resulting in part from leakage through the waterproof membrane (Porter 1985). Also, leakage of de-icing salt is believed to have contributed to the pitting corrosion which led to the sudden collapse of a post-tensioned bridge deck at Ynsygwas (Middleboe, 1985).

Unwaterproofed decks subject to de-icing salt can undoubtedly be expected to be deteriorating. This problem first manifested itself in the USA where membranes were not used, with the result that damage estimated at several billion dollars has occurred. In Britain, the deck of Barton Bridge (Fig. 8.2), which was largely unwaterproofed, has had to be replaced (Wood, 1986). In addition to the problems arising from salt on the top surface of decks, the soffits of overbridges are vulnerable to the effects of salt spray.

Very few bridge deck expansion joints are completely watertight and hence de-icing salt can run down the ends of decks which are virtually never

Fig. 8.2. Barton Bridge.

protected from chloride ingress. In reinforced concrete decks this can lead to corrosion at the deck ends, which is serious enough; however the risk posed to tendon anchorages at the ends of post-tensioned prestressed bridges might be unacceptable if the tendons are not properly grouted.

Interestingly, concrete parapets on bridges do seem to be resilient to the effects of de-icing salt despite being exposed to splash and spray from traffic. Although problems have been experienced on some bridges, most concrete parapets perform well. Parapets will readily shed any contaminated water but more importantly they are exposed to the beneficial cleansing action of the rain.

8.2.2 Bridge substructures

Leakage through deck joints carries de-icing salt to the tops of piers and abutment-bearing shelves. Often this water then runs down the face of the pier or abutment. On overbridges this chloride contamination is exacerbated by spray. The surfaces of substructures are rarely waterproofed and are sheltered from rain. In consequence, substructures are likely to be the most vulnerable elements of bridges as far as chloride-induced corrosion is concerned.

Fig. 8.3 Midlands Links Viaduct substructure.

Many instances of chloride-induced corrosion of substructures have been reported. For example, chloride contamination of the substructures of the viaducts on the Midland Links motorways is widespread (Fig. 8.3) and at Camsley Lane Viaduct £200 000 has been spent repairing extensive delamination and cracking in five crossheads (Ambrose, 1985).

Substructures are particularly vulnerable in areas close to ground level where there is the opportunity for salt to soak into backfill. This tends to result in increased chloride contamination and the generally damp conditions which exist below ground level will encourage corrosion.

8.2.3 Retaining walls

Roadside retaining walls appear to fare much better than bridge substructures. This is probably a result of the beneficial action of rain in cleansing the surface of concrete.

8.3 Diagnosis

Failure to carry out a proper examination of a structure and correctly diagnose the cause of deterioration could mean that repairs are incorrectly specified. The likely consequences of this are that repairs will fail very quickly and abortive expenditure will be incurred.

A visual survey will provide valuable information on where deterioration is taking place and give some indication of the likely cause. This survey needs to be supplemented by testing to confirm the cause, extent and, if possible rate of deterioration.

The use of a potential mapping technique for the detection of sites of active corrosion was first promulgated in the USA in the 1970s. The technique is now widely used in the UK and provides a simple, efficient and inexpensive means of locating corroding reinforcement. Care does need to be exercised, however, in the interpretation of the data. The only published standard for the technique (ASTM, 1987) gives guidance on interpretation based on data derived from work carried out on bridge decks in the USA. The guidance may not be universally applicable to all concrete structures and, in addition, corrosion potentials are modified by a number of factors, particularly ambient conditions (Figg and Marsden, 1985). Instances have been observed where potentials have shifted by more than 100 mV during the course of 24 hours as a result of humidity changes. Half-cell potential measurements should therefore only be entrusted to experts who are fully aware of the pitfalls associated with the technique. Such people should be able to provide reasonably accurate assessments of the likelihood of corrosion and, with experience, it is even possible to determine with some accuracy whether general corrosion or pitting corrosion is occurring.

Having determined that corrosion is taking place the next requirement is to assess how much section has been lost from the reinforcement or prestressing steel. Also, because it is rarely possible to commence repairs immediately, it would be very useful, if not essential, to have some appreciation of corrosion rate.

As yet there are no practical non-destructive test methods available for determining how much section has been lost from reinforcement as a result of corrosion. Radiographs have been obtained from reinforced concrete members using a linear accelerator to generate a powerful X-ray beam. The results were inconclusive and the cost is likely to militate against the use of the technique as an everyday test procedure. However, it may prove to be useful for limited special applications such as the examination of vulnerable areas of prestressing tendons contained within ducts. General corrosion usually leads to spalling of concrete; hence the facility to measure loss of section is readily available. Pitting corrosion, such as can occur in the presence of chlorides, can lead to severe local loss of section, sometimes without there being any visual evidence. If pitting corrosion is suspected, therefore, it is essential that its effect on structural integrity is assessed. At the moment, this can only be done by exposing corroding reinforcement and measuring section loss.

Concrete resistivity measurements provide an indirect indication of corrosion rate. Concrete resistivity is related mainly to moisture content; hence changes in ambient conditions can have a marked effect on resistivity. As in the case of half-cell potential testing, the carrying out and interpretation of resistivity measurements is best entrusted to experts. Other means of measuring corrosion rate such as electrochemical noise and electromagnetic flux measurements are in the course of laboratory development but are not yet available for site use.

Having determined that corrosion is taking place, the cause can be readily investigated. The use of phenolphthalein as an indiator to measure depth of carbonation is well documented as is the removal of concrete dust samples for subsequent laboratory determination of chloride content. These measurements need to be allied to concrete cover measurements to determine whether the carbonation front or chloride ingress has progressed to the level of the steel. All these test techniques have been discussed in some detail in Chapter 3, section 3.3.1.

8.4 Options for repair

Where corrosion is identified in a highway structure there are six procedures which can be considered:

1. do nothing;
2. reduce corrosion rate;
3. repair visual defects;
4. carry out major repairs;
5. apply cathodic protection;
6. replace affected elements.

The approach taken to repairs will almost certainly be influenced by cost. Clients will quite rightly expect structures to be repaired using the most cost-effective techniques. In this context the cost of maintenance subsequent to repairs must be considered in addition to the initial repair costs.

Unfortunately, there are no hard and fast rules as to which will be the most economic repair to adopt in any given circumstance. Obviously, it will be possible to predict the likely cost of a repair procedure but, as yet, there is little information available on durability of repairs, particularly where they are necessitated by chloride-induced corrosion. The size of the structure under consideration will perhaps be an important influence on the choice of repair procedure. It will be far easier to embark on repairs to a small structure, such as a retaining wall, where the financial risk associated with not quite getting the repairs right are small compared with those arising from implementing the wrong repairs to a multi-million pound motorway structure.

8.4.1 Do nothing

If the appearance of a highway structure is not of primary concern and spalling concrete does not pose a risk to the public, it may be possible to consider not carrying out repairs to a structure. Before this option could be exercised a structural assessment would need to be carried out to confirm that the structure has a reserve of strength in its corroded condition. Most importantly, the structure would need to be monitored to ensure that continuing corrosion does not jeopardise structural integrity. However, as there are no methods currently available for measuring loss of reinforcement cross-section it would be difficult to judge the point at which a structure is no longer safe. While this risk might be acceptable in a reinforced concrete structure affected by general corrosion which might be expected to provide ample warning of distress, it would be completely untenable in a prestressed concrete deck where failure could be sudden and catastrophic.

8.4.2 Reduce corrosion rate

The application of a coating to chloride-contaminated concrete has been advocated as a means of reducing, or even arresting, corrosion. This is on the premise that a coating will allow the concrete to dry out by vapour transference, thus removing one of the ingredients essential for corrosion. Similarly, complete encapsulation of concrete members has been promoted as a means of starving the corrosion cells of oxygen, thus inhibiting the corrosion process.

It is considered unlikely that either approach will work (Chapter 5, section 5.10). Coatings capable of allowing the passage of water vapour from concrete to the atmosphere are equally capable of allowing vapour to move in the opposite direction. The hygroscopic properties of salt are almost bound to ensure that this happens. It is therefore likely that the moisture content of coated concrete members will remain at or about ambient relative humidity, which in Britain is high enough to promote corrosion throughout long periods of the year. If concrete is encapsulated it is probable that there will be sufficient oxygen within the concrete to maintain corrosion. There is also the danger that depleting the oxygen content could give rise to the conditions necessary for pitting corrosion.

8.4.3 Repair visual defects

Where visual defects are attributable to carbonation-induced corrosion of reinforcement an approach involving making good only those areas seen to be defective can be a viable repair procedure. Corrosion is usually initiated on reinforcement having the least cover as this is the steel that the carbonation front reaches first. Thus the number of corrosion sites is usually limited and, if the steel is repassivated by the application of an alkaline repair material, then corrosion can be arrested. There are many proprietary systems available for the repair of carbonation-induced corrosion and properly used they can be used to effect durable repairs (Chapter 4). The systems usually comprise a polymer-modified mortar with good carbonation resistance which obviates the need for restoring full concrete cover to the steel. It is usual to coat the repair areas as well as adjacent areas of concrete with a carbon-dioxide-resistant coating to prevent further carbonation. Proprietary repair systems usually also include a rust inhibitor that is applied to the reinforcement prior to repairing the concrete. The reason for this is unclear because the repair mortar is quite capable of repassivating the reinforcement itself.

Making good areas which are seen to be defective by way of cracking, spalling, delaminations etc. should only be considered as a short-term remedy if the corrosion which has caused the damage has resulted from the ingress of chlorides. Inevitably, chlorides will remain in areas of concrete surrounding the repair but will not have stimulated corrosion because of cathodic protection afforded by adjacent corroding anodic zones. Once repairs have been carried out, there is nothing to prevent depassivation of the reinforcement in the non-repaired areas and corrosion will be stimulated by the cathodes formed in the repaired areas. This corrosion will very quickly lead to further deterioration which is likely to occur initially around the boundaries of repaired areas (Chapter 5, section 5.3).

8.4.4 Carry out major repairs

Repairing structures by removing chloride-contaminated concrete and replacing it with fresh concrete is much more likely to succeed than simply making good visual defects. There are a number of important questions that need to be addressed, however, before such a repair procedure is embarked upon.

First, it will be necessary to determine how much concrete has to be removed and this will require the appropriate chloride threshold to be established. For reasons of cement chemistry the level of chloride necessary to initiate corrosion is different according to whether it was present at the time of mixing the concrete or entered the concrete after it had hardened. The presence of carbonated concrete will also influence the level of chloride necessary for corrosion to occur (Neville, 1983). There is no consensus as to what the chloride threshold is, or in fact whether there is any such thing as a threshold. As far as the repair of concrete members affected by chloride which has entered the concrete post-hydration is concerned, it is suggested that the removal of all concrete having a chloride content of 0.2% chloride ion or more by weight of cement should provide a high probability of success (Vassie, 1984).

The accurate measurement of the chloride content of concrete is notoriously difficult. Chloride content determinations are normally made in the laboratory by means of a titration analysis. While the laboratory technique is reasonably accurate, great care needs to be exercised in the collection of samples on site. Samples usually comprise dust obtained from drilling into the concrete. Sampling equipment has to be scrupulously cleaned between samples to avoid any contamination and even when this is done there is plenty of opportunity for sampling error. The chloride is contained in a cementitious fraction of the concrete which means that different results will be obtained according to how much aggregate is incorporated in the samples. In addition, chloride thresholds tend to be expressed with respect to cement content which is very difficult to measure accurately from a concrete sample. As-built records are a more reliable source of cement content information.

Having established how much concrete to remove, the means of breaking out concrete has to be specified. The concrete to be removed is often sound and of high quality and removal has to be carried out without causing damage to the reinforcement or the remaining concrete. While pneumatic tools have traditionally been employed for the removal of concrete, they can easily damage reinforcement and sometimes cut right through smaller diameter bars. Pneumatic tools can cause microcracking of the concrete, and aggregate which is shattered and loose is often left in the surface against which the repair concrete is to be placed.

In contrast, high pressure water jetting can remove concrete without damaging the remaining concrete or reinforcement. It will also leave the concrete surface in an ideal condition to accept the repair concrete. Water jetting will also remove corrosion products from the reinforcement and thus obviate the need for the second-stage operation of cleaning reinforcement that is associated with mechanical break out.

High pressure water jetting does have the disadvantage that it is considerably more expensive than the use of mechanical tools. With equipment currently available the operation is also significantly slower and it is difficult to control. The reasons for this are twofold: first, the reaction forces associated with discharge of water at pressures of up to 900 bar and 90 litres per minute mean that the manual use of the equipment is very tiring; second, the amount of water used makes it very difficult for the operative to see how the cutting operation is progressing. The reaction problem can be overcome by the use of rig-mounted jetting lances but these tend to restrict the usefulness of the equipment. Hand-held lances are still required for final trimmings and for the removal of concrete from behind reinforcement. The drawbacks of the water jetting equipment currently available may be overcome by improved pumps and ancillary equipment soon to be introduced in this country. The pumps operate at much higher pressures but with lower water output – 2000 bar and 12 litres per minute respectively. Thrust on the lance is reduced to approximately 12 kg and therefore the equipment can be comfortably hand held, the lower volume of water gives the operative a better opportunity of seeing what he is doing, and the higher pressure means that concrete should be removed faster than with equipment currently available.

The final step in the repair process is to replace the concrete. It is unlikely that conventional concretes could be used if for no other reason than that they will almost certainly crack due to restraint to shrinkage provided by the concrete surrounding the repair area. Usually, the total volume of concrete involved in repair work will be small and the volume incorporated in any one repair area will not be an attractive proposition to a ready-mixed concrete supplier. Maintaining adequate quality control over all volumes of concrete batched on site will also prove difficult. These factors tend to favour the use of proprietary shrinkage-compensated repair concretes where all the contractor has to do is add water to a pre-bagged material. Such concretes will almost certainly have to be used for difficult repairs such as those to the soffits of beams with heavily congested reinforcement where a shrinkage-compensated concrete having good flow properties is required. For further information on concrete repair materials and techniques refer to Chapter 4.

This approach to repairs will usually require the removal and replacement of significant amounts of concrete and is therefore almost certain to necessitate the use of temporary support to structural members even if traffic can be temporarily diverted off the structure. Where traffic loads have to be carried by a structure under repair, the repairs will have to be carried out in a piecemeal fashion to ensure that there is always a path through the structure for live loads.

Where chloride contamination is extensive, the high costs of temporary support, concrete removal and replacement may mean that the repair procedure is more expensive than the complete replacement of defective elements. Even where the procedure is apparently cost effective, the durability of the repairs will be unknown. In this respect it is uncertain whether all chloride contamination can be removed from reinforcement, particularly from within pits. The structural effects in terms of redistribution of stresses during the course of repairs are also unknown, but see Chapter 4.

8.4.5 Apply cathodic protection

Since the corrosion of steel in concrete is an electrochemical process, electro-chemistry can be used to provide a means of mitigating corrosion. Cathodic protection is achieved by impressing a low-voltage direct current from an anode system placed on the concrete surface through the concrete and onto the steel. The cathodic protection current opposes the current associated with the corrosion process. When sufficient current flow is achieved the corrosion current will be suppressed (Chapter 5; Concrete Society, 1989).

Making good any areas of spalled or delaminated concrete are the only concrete repairs required prior to the installation of an impressed current cathodic protection system. Problems associated with incipient corrosion at the boundaries of repaired areas that are inherent with simply repairing damaged areas are prevented by the cathodic protection current. Thus, cathodic protection has a marked advantage over other procedures in that the need to install temporary support systems is avoided provided that the loss of section due to corrosion has not impaired structural integrity. The cost of carrying out any necessary concrete repairs and installing a cathodic protection system on a structural element with extensive chloride contamination will be significantly less than either replacing or repairing the member (Boam, 1987). The application of cathodic protection is therefore an attractive proposition for the rehabilitation of elements of highway structures affected by chloride-induced corrosion.

In North America cathodic protection is recognized as one of the best means of rehabilitating reinforced concrete bridges affected by chloride-induced corrosion (Hay and Virmani, 1985). The procedure has been in use for over 14 years in the USA and Canada where it was initially applied to bridge decks. However during the last four or five years it has also been used on substructures. While some information is available from the systems installed in North America, there are still no quantitative data on the life of anode systems or on maintenance costs. This makes it difficult to select the most appropriate anode system for a given installation to estimate the total cost of specifying cathodic protection as a rehabilitation procedure.

There is also, perhaps understandably, a reluctance among many specifiers to adopt cathodic protection for the repair of structures. Conventional concrete repairs present tangible evidence that the corrosion affecting a structure has been overcome and will be regarded by most engineers as a once-and-for-all solution. (It remains to be seen, however, whether or not this is the case.) In contrast, electrochemistry is an alien field to the majority of structure owners as well as to civil and structural engineers. To rely on such a remedy to provide a cure for a deteriorating structure is much harder to accept. Also, the finite life of anode systems at present available is small compared with the 120 year design life of bridges in Britain and the need for future maintenance also tends to militate against cathodic protection. It is therefore probable that, although cathodic protection may well be the most economic approach to the rehabilitation of highway structures affected by chloride-induced corrosion, its widespread adoption will be delayed until laboratory and field trials of the procedure which are currently in hand have been completed.

8.4.6 Replace affected elements

The demands on the road network in Britain are such that it will rarely be possible to close structures in order to facilitate the replacement of defective elements. In the case of motorways even lane closures can give rise to serious traffic disruption with its significant concomitant costs. Replacement of elements of bridge structures would therefore almost certainly necessitate the use of temporary support systems to carry dead and live loads while the work is carried out. Such temporary support systems are likely to be very expensive. However, by completely replacing defective elements it would be possible to obtain a structure with good durability provided, of course, that measures are implemented to avoid a recurrence of the conditions that led to the initial deterioration.

Not all the repair options discussed above are applicable to prestressed concrete structures. Such structures are not amenable to procedures involving the removal of concrete unless the repairs are of a minor nature such as those necessitated by the carbonation-induced corrosion of secondary steel. It is likely that the only viable repair option currently available will be to replace the defective elements or to incorporate permanent additional support. Cathodic protection is currently not recommended for application to prestressed concrete because there is a possibility that hydrogen embrittlement of the prestressing steel occurs at certain levels of protection. Also, the cathodic protection current causes sodium and potassium ions to move towards the cathode, i.e. the prestressing steel. There is also some concern that this could have an effect on the bond between the prestressing steel and the concrete. Research on the application of cathodic protection to prestressed concrete is being carried out in North America with some encouraging results. There may therefore be some scope for this procedure in the future.

Cathodic protection is also not recommended for application to carbonated concrete. This is because carbonation increases the resistivity of the concrete, making it more difficult to impress an electric current. In any case, the usually limited extent of carbonation-induced corrosion is unlikely to warrant this procedure and repairs using proprietary mortars etc. are likely to be more cost effective.

The foregoing repair options are not mutually exclusive and could be used in combination. This could be useful when dealing with large and complex structures where funds and resources will not usually be available to carry out all repairs simultaneously. A cost-effective repair strategy for such structures could be developed by looking at different combinations of options. For example, the use of cathodic protection could be contemplated to prevent further deterioration of a particular element or series of elements while more radical solutions were implemented in other areas of the structure; it may also be appropriate not to carry out repairs to defects elsewhere in the structure. This round of repairs, or inactivity, could be followed in time by, say, the replacement of cathodic protected elements when the anode systems reach the end of their useful lives.

By programming and costing all combinations of options for repairs according to the needs of the structure and funds and resources available, it will be possible to identify the most cost-effective maintenance strategy appropriate to the structure.

8.5 Monitoring of repairs

While the options available for the repair of highway structures are described in this chapter, little is known of the long-term performance of repairs, particularly those used to mitigate effects of chloride-induced corrosion. For the 'one-off' relatively small repair it is probably unnecessary to do more than carry out a visual inspection of the structure from time to time. However, where large-scale repairs are carried out to major structures as part of an ongoing repair programme it is vital to have an early appreciation of the effectiveness, or otherwise, of the repairs. Only in this way can the compounding of mistakes be avoided and repair techniques be improved.

Where concrete repairs are implemented to arrest corrosion it is obviously important to determine that corrosion has stopped. Currently, the only means of assessing corrosion activity non-destructively is to carry out half-cell potential measurements. However, as potentials change with ambient conditions a false impression could be gained by taking only a single set of potentials after repair. Therefore, at least two or three sets of potential readings will be needed to be sure of obtaining an accurate interpretation of the data. Unfortunately, it is difficult to detect subtle changes from equipotential plots, which is the normal way of presenting potential data, and it is nigh on impossible to identify what is happening by looking at the raw data. Changes in potential can be assimulated much more easily if the data are presented in a cumulative frequency plot as shown in Fig. 8.4.

The advantage of the cumulative frequency diagram is that changes in the corrosion behaviour of a structure can be readily identified. For example, if corrosion has been mitigated by a repair procedure the cumulative frequency curve should move to the left of the plot representing potentials measured prior to repair. If corrosion has stopped altogether, then the curve would ideally be wholly in the range 0 to $-200\,mV$. By also inputting where on a particular structure various potentials have been measured, it is also possible

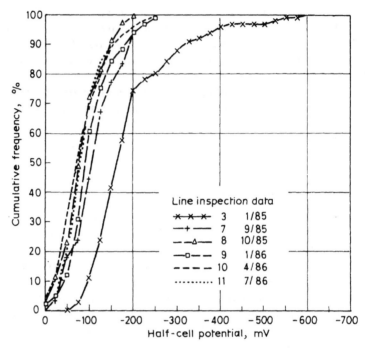

Fig. 8.4 Half-cell potential cumulative frequency plot for motorway crossbeam

to use the programme to compare the performance of different areas of the structure. Similarly, the performance of repaired areas can be compared with that of non-repaired areas.

The plots in Fig. 8.4 represent six sets of data selected from a series taken as part of a regular half-cell potential monitoring exercise on a crossbeam supporting a motorway viaduct. Inspection 3 made in January 1985 was carried out prior to repairing the beam during the course of the summer. All the other inspections represented were made subsequent to repairs. It can readily be seen that the repairs have effected an improvement measured in terms of half-cell potential activity.

8.6 Avoiding future problems

Much can be done to improve the durability of new British bridges and other highway structures.

Prohibiting the use of salt for de-icing would obviously avoid the problems caused by chloride. Urea, which costs approximately ten times as much as salt, is being used on the Severn and Avonmouth bridges as well as on the Midlands Links Viaducts. The additional costs can be justified for these structures which are of significant economic value and contained in compact areas of the highway network. It is unlikely that the additional de-icing costs will be acceptable for other roads in Britain where on average there are only three or four bridges per kilometre. Calcium magnesium acetate which is a more efficient de-icer than urea has also been suggested as an alternative to salt. However, this material costs 40 times as much as salt.

As far as deck slabs are concerned, spray-applied plastic waterproof membranes now coming into use offer advantages over some of the earlier waterproofing systems. These new membranes are fully bonded to the concrete; hence any punctures are unlikely to allow significant penetration of chlorides into decks. Also, it is much easier to seal these membranes around projections such as manholes etc.

Minimizing the number of expansion joints in decks by designing continuous structures should ensure a marked improvement in the durability of substructures. Where joints have to be used, the provision of positive drainage below them would avoid contamination of the substructures.

Concrete surfaces exposed to possible salt penetration can be waterproofed by the use of surface treatments, some of which can reduce the ingress of moisture and chlorides by up to 90%.

The use of epoxy-coated reinforcement in areas of bridges that are vulnerable to salt penetration is mandatory in Ontario, Canada. Properly used, coated reinforcement, which is now available in Britain, can provide very good durabilty.

Improvements to concrete mix design can also significantly improve the durability of structures. Current codes of practice specify lower water–cement ratios and higher cement contents than previously. Both these factors will provide the concrete with increased resistance to chloride penetration. Even greater improvements can be achieved by the inclusion of cement replacement materials such as ground granulated blastfurnace slag, microsilica and pulverized fuel ash. These materials modify the pore structure of the concrete, increasing resistance to chloride penetration by 10–20 times.

None of the measures described above which are designed to improve durabilty will be of any use if high standards of workmanship and good quality control are not exercised during construction. If concrete is inadequately cured then durability will suffer markedly, particularly where cement replacement materials are used. Even the most impervious of concretes will be useless if reinforcement is fixed with low cover. Damage to epoxy coatings caused by careless handling may enhance the corrosion of damaged areas.

All too often the owners of highway structures have focused their attention on the initial cost of construction and not considered the costs involved in maintaining their structures. It is now becoming all too apparent that maintenance costs are a significant consideration – in some instances the costs of repairs to structures are going to be much more than it cost to build the structures in the first place. The decision to replace many of the worst affected crossheads on the Midland Links Viaducts rather than to carry out *in situ* repairs (Watson, 1990) is evidence of the scale of costs involved. Based on the results of site testing, the costs of repair and maintenance have been calculated for the 200 bridges in the Department of Transport Study (1989) and extrapolated to the 5933 concrete bridges owned by the Department. The total cost of carrying out the recommended remedial work on all the bridges would amount to about £620 million over ten years.

Any measure designed to improve the durabilty of a highway structure is bound to add to initial costs. In view of the current problems, it is suggested that in the long term this will be regarded as money well spent.

References

Ambrose, K. (1985) Chloride contamination of Camsley Lane Viaduct, *Construction Repairs and Maintenance*, September, pp. 7–9.

ASTM (1987) *Standard Test Method for Half-Cell Potentials of Uncoated Reinforcing Steel in Concrete: ASTM C876–87*, ASTM Annual Book of Standards, vol. 4.02.

Boam, K.J. (1987) The application of cathodic protection to highway structures – the consulting engineer's viewpoint, Paper presented at *Seminar on Corrosion in Concrete – Practical Aspects of Control by Cathodic Protection*, May, London.

Concrete Society (1989) *Cathodic protection of reinforced concrete*, Report of a Working Party, *Technical Report 36*, Concrete Society and Corrosion Engineering Association.

Department of Transport (1989) *The Performance of Concrete in Bridges. A Survey of 200 Highway Bridges*, HMSO, London, April.

Figg, J.W. and Marsden, A.F. (1985) *Development of Inspection Techniques for Reinforced Concrete: A State of the Art Survey of Electrical Potential and Resistivity Measurements for Use above Water Level*, Concrete in the Oceans Technical Report 10, London.

Hay, R.E. and Virmani, Y.P. (1985) North American experience in concrete bridge deterioration and maintenance, Paper presented at *Symposium on Concrete Bridges – Investigation, Maintenance and Repair*, September, Concrete Society, London.

Institution of Civil Engineers (1988) *Report on the State of Roads and Bridges in the United Kingdom – the Case for Action*, January.

Middleboe, S. (1985) Welsh bridge failure starts tendon scare, *New Civil Engineer*, 12 December, p. 5.

Neville, A (1983) Corrosion of reinforcement, *Concrete*, **17** (6), 48–50.

Porter, M.G. (1985) Repair of post-tensioned concrete structures, Paper presented at *Symposium on Concrete Bridges – Investigation, Maintenance and Repair*, September, Concrete Society, London.

Vassie, P.R. (1984) Reinforcement corrosion and the durability of concrete bridges, *Proc. Inst. Civil Eng., Part 1*, **76**, August, pp.713–732.

Watson, R. (1990) Spaghetti Junction lifts off £400m repairs, *New Civil Engineer*, 28 June, p. 7.

Wood, J.G.M. (1986) Refurbishment and renovation of bridges, *Structural Engineer, London*, **64A** (2), 57–5CS2.

Railway structures 9

J.F. Dixon
(Formerly British Rail Research)

9.1 Introduction

The bridge is the most common railway structure and around the railway network of Britain there are currently 38 000 of them. Many are constructed of brick and masonry, some of iron and steel and a few of timber. Of course, a fair number are made principally of concrete; in fact there are 3800 constructed from a wide variety of concrete types. From the historic and spectacular Glenfinnon Viaduct (Fig. 9.1) to the modern and controversial cable stay bridge at Virginia Water in West London (Fig.9.2) the range is wide. Other significant railway structures include office and station complexes in most British cities, a number of sea defences and harbour installations, and naturally, track components.

With such a variety of concrete structures spread over almost all parts of Britain, exposure to many types of environment occurs. It is perhaps not surprising, therefore, that British Rail civil engineers have between them encountered all the known concrete deterioration mechanisms — from steel corrosion to alkali–silica reaction, from sulphate to frost attack. There is no doubt that the experience gained has caused engineers in lots of instances to be dissatisfied by the lack of durability achieved. Many were originally of the opinion that concrete was maintenance free and budgets have been stretched by the need to repair or replace prematurely. The main problem encountered has been that of early steel corrosion due to chlorides or carbonation or both.

The need to check so many structures led British Rail Research to develop a portable test kit for on-site testing. This carries the equipment and instructions necessary to permit a rapid assessment of chloride levels and carbonation depths (Fig. 9.3). It is now being used throughout the British Rail network by local engineers (Child, 1988).

Some recent British railways cases where concrete deterioration has presented problems are discussed in this chapter.

Fig. 9.1 Glenfinnon Viaduct.

9.2 Steel reinforcement corrosion

9.2.1 Chloride problems

Bridge 84, Newport Road, Stafford

This first case study describes a problem caused by water containing chlorides leaking through a bridge deck and then running down the faces of concrete

Fig. 9.2 Virginia Water Cable Stay Bridge.

beams and columns. The bridge, shown in Fig. 9.4, carries the A518 road traffic over the west coast main line track just south of Stafford railway station. Constructed in 1961, the bridge has four spans of pre-stressed concrete deck beams supported on reinforced concrete trestle piers.

Unfortunately at the time of construction the deck was not waterproofed and so water has seeped through the deck joints and down the faces of the pier crossheads and columns (Fig. 9.5). During the winter months this water often carries with it chlorides from local authority road salt applications.

Fig. 9.3 British Rail Research site test kit.

A belated attempt to stem the contamination was made in 1984 by waterproofing the deck. By then, however, much of the damaging chloride penetration had occurred. As has been explained in Chapter 2, section 2.2.4, if a sufficient quantity of the chloride reaches the vicinity of the reinforcing steel some corrosion is likely. With some of the columns of bridge 84 the cracking and resultant spalling were spectacular (Fig. 9.6).

In an endeavour to assess the extent of non-visible corrosion, half-cell and chloride surveys were conducted. These not only confirmed the expected high chloride content (up to 0.25% by weight of sample in places) but demonstrated too that the depth of contamination was deep and wide-spread. Chloride profiles along cores cut from a crosshead beam are shown in Fig. 9.7.

The most surprising and worrying feature of the half-cell survey was that potential values of around −500 mV were spread over the whole of one crosshead beam. Confidence in these results was boosted, though, when it was shown that the accepted 'critical' levels of below −350 mV had been

Fig. 9.4 Bridge at Newport Road, Stafford.

exceeded only in areas of concrete clearly stained by the water leakage. A typical 250 mm grid plot of readings from one column face is shown in Fig. 9.8. The deterioration rate had no doubt been boosted by deep carbonation penetration. Depths of up to 40 mm were measured by the phenolphthalein test.

Following the investigation, much debate ensued on the wisdom of carrying out repairs. The devastating penetration of chloride meant that concrete containing more than the then acknowledged critical content of 0.45% by weight of cement would remain, short of complete demolition of components. It would not be easy to predict the life of a repair and the prospects of cathodic protection were only just being examined on a structure elsewhere in the network. The Midland Region civil engineers opted to repair, aware that it was merely a holding operation and that future replacement would need to be planned.

The repair system chosen consisted of the basic preparation procedure of the removal of all cracked and spalled concrete. In some areas this meant removing up to 200 mm depth of concrete, thus exposing inner bars. When steel was exposed it was specified that concrete some 25 mm behind it

Fig. 9.5 Evidence of water seepage and spalling of columns, Stafford Bridge.

should go too. Structural considerations dictated that neighbouring columns could not be repaired concurrently.

All exposed corroded steel was grit blasted and treated with two coats of an epoxy resin priming coat dashed with quartz sand. The one chosen contained cement clinker which, the manufacturer claimed, mops up moisture. The validity of such claims has been commented on earlier in Chapter 4, section 4.2.5. Where more than 60% of the steel cross-section had been lost, replacement reinforcement was added.

Replacement depths of greater than 25 mm were poured behind shuttering, but with thinner sections the acrylate-copolymer-modified repair mix was

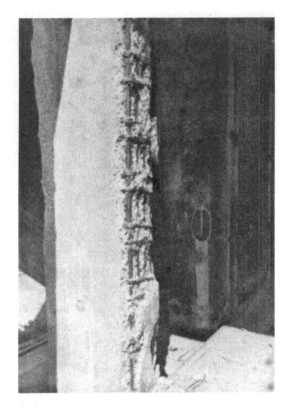

Fig. 9.6 Corrosion damage to column, Stafford Bridge.

hand applied. The final treatment chosen was a coating of a hydrophobic impregnated siloxane followed by a primer and then two coats of an elastic acrylic top coat to give a white finish. The elastic coat was chosen because of its alleged crack-spanning properties. Figure 9.9 shows the complete repair of one column.

It is reported that some two and a half years after application all is sound apart from a few ominous rust stains.

Bridge 42, Scunthorpe

This is another example of a bridge which suffered because water percolated through the road deck. In this case, however, demolition was the chosen solution for the 56 year old structure (Fig. 9.10) replaced in 1984.

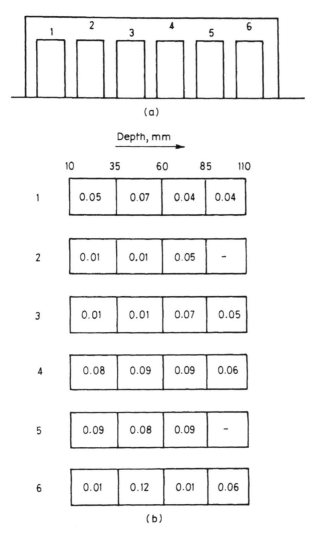

Fig. 9.7 Chloride distribution, crosshead beam, Stafford Bridge:
(a) core location; (b) chloride profiles (per cent by weight).

The concrete in this bridge was considered unique for two reasons — first, for the exceptionally high chloride levels, up to 1% by weight of sample in parts, and second for the extremes of carbonation measured. Depths varied from 40 mm down to just 1 mm. The bridge was of *in situ* concrete and so

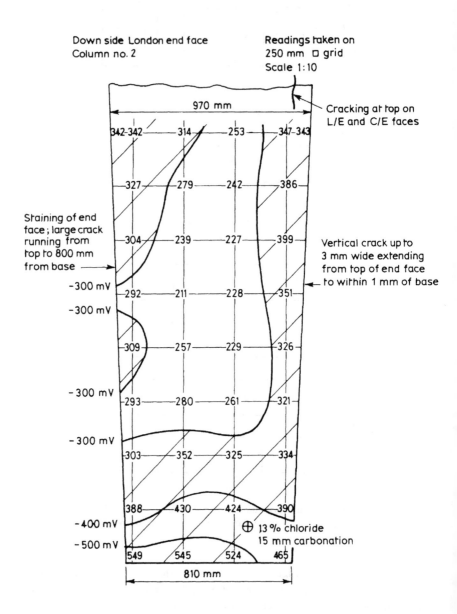

Fig. 9.8 Half-cell plot, column face, Stafford Bridge.

Fig. 9.9 Repaired column, Stafford Bridge.

variation could be expected, but it was interesting to note that the areas with low carbonation were coated with soot.

Underbridge at Bassaleg, near Newport, South Wales

The bridge (Fig. 9.11) carries the Western Valley line across the M42 just outside Newport. It was completed in 1964 some three years before the motorway was opened to road traffic. Inevitably during winter months when de-icing salts are applied to the road surface, fast-moving traffic in the outer lanes liberally sprays the central tapered pier with a chloride solution. Compaction

Fig. 9.10 Bridge 42, Scunthorpe.

Fig. 9.11 Bassaleg Bridge, Newport, South Wales.

of the *in situ* concrete was poor in some places and cover to reinforcement low. An early sprayed concrete 'repair' was applied to compensate for this when the bridge was three years old. Alas this was not enough and soon the salt began to take its toll. In 1984 some cracking in the pier was noted in an inspector's report. It was described as severe and a diagnostic survey was called for.

Chloride measurements and a half-cell survey were carried out on both longitudinal faces of the central pier. A summary of the work on the south elevation (Fig. 9.12) hints at a relationship between the two sets of results. Whatever the link, the chloride levels were high enough to have initiated the corrosion and the cracking and the half-cell results confirm that hidden corrosion could herald more trouble.

Access to work in the central reservation of a motorway is expensive and troublesome and so any remedial measures needed to be durable. The Board agreed that the situation was an ideal one to establish the merits of a cathodically protected repair. This could prove to be a milestone for bridge repair and engineers both within and outside British Rail would watch developments with interest.

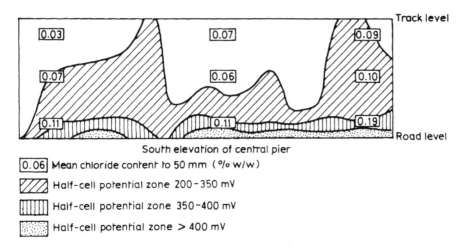

Fig. 9.12 Chloride and half-cell results, Bassaleg Bridge.

Use of calcium chloride accelerating admixture

The third type of chloride problem arose from its use as a concrete-accelerating admixture. Now prohibited in concrete containing steel, its use during the 1950s and 1960s was common. Indeed, proprietary cements known as extra rapid-hardening varieties were available in those days. They contained 1.5%–2% of calcium chloride.

The British Railways Board gained unwanted but subsequently valuable experience with the corrosion problems that the use of such materials can cause. In 1963 the Board leased a multistorey office block in London. It was constructed using precast reinforced concrete beam and column units containing extra rapid-hardening Portland cement, and after only six years cracking was reported. Subsequent actions mirror the evolution of deterioration mechanisms and knowledge of concrete repair. The prevalent view at the time was that the cracks should be filled with epoxy resin. It was thought that the injection of resin, because of its strength, would stop cracks developing and its protective qualities would prevent further corrosion. Most of the cracks were thus duly injected.

It did not take many months for this action to be proved futile. The cracks continued to open and new ones appeared. It was also revealed that at the time the importance of resin viscosity was not appreciated. The resin in many cases had merely drained away down the cracks without bonding the faces together.

The injection of cracks with resin does still have an established role in concrete repair but the causes of cracking and the relevance of resin rheology are now better understood (Chapter 4, sections 4.2.6 and 4.3.7).

With this case, more thorough investigations followed and by now the damaging effects of chloride were being recognized. The concrete, it was found, was of poor quality and carbonation, up to 40 mm in places, added to the problem. It was now realized that a long-term repair was not possible and so a repair design was sought which would at least retard the rate of deterioration. Calculations showed that the concrete could not be removed from all around the steel in the damaged areas without considerable propping to maintain the structural integrity of the building.

Little published guidance was available in the early 1970s, but the sensible approach of cleaning exposed steel and avoiding feather edges was adopted. The specification called for a cementitious patch repair mix modified with a styrene–butadiene polymer. These polymers, it was claimed, improved the bond and the durability (Chapter 4, section 4.2.5). It was further claimed (but nowhere published) that bond to the original concrete could be enhanced if a bond coat consisting simply of Portland cement and styrene–butadiene rubber, plus perhaps some sand, was painted onto the parent concrete ahead of the repair mix. Such a coating was used in this repair. Alas, after less than a year the repairs began to fail (Fig. 9.13) and subsequent investigations pointed to the bond coat as the prime cause of premature failure. This led to some debate regarding the efficacy of bond coats (Dixon and Sunley, 1983) but since then a good deal of research has led to revised application techniques and increased confidence in their use.

Fig. 9.13 Repair failure on multistorey office block, London.

9.2.2. Carbonation

Carbonation alone has not proved to be a major problem with concrete on British Rail structures. Some decorative cladding panels at Banbury Station (Fig. 9.14) are a rare example of carbonation-induced reinforcement corrosion. Here, although the decorative limestone coarse aggregate particles were held in an extremely weak cementitious mortar reinforced with a steel mesh, the carbonation front had penetrated fairly rapidly beyond the steel and the subsequent corrosion had caused the unsightly spalling shown. Once again repair was considered to be inappropriate and the panels were replaced.

Fig. 9.14 Carbonation at Banbury Station.

9.3 Concrete deterioration

9.3.1 Frost attack

Frost attack in non-air-entrained concrete has claimed or damaged some surprising victims. Prestressed concrete railway sleepers made from concrete of grade 75 N/mm^2 have been known to succumb (Fig. 9.15). This is quite rare, though, and only occurred in cases where the aggregate quality was in question. However, it does raise doubts about the general assertion that high strength necessarily guarantees a durable concrete.

Fig. 9.15 Frost attack, prestressed concrete sleeper.

Another notable victim of frost has been some of the columns of the historic Glenfinnon Viaduct. Majestically skirting the northeast shore of Loch Shiel, this curved *in situ* unreinforced concrete structure built towards the end of the last century has generally weathered well. Unfortunately the deck was not originally waterproofed and so column bases have been regularly saturated as water has soaked freely down the lean concrete in the core of the columns. The outer concrete has thus been subjected to disruption during freezing, leading to some fairly severe spalling (Fig. 9.16). Simple patch repairs have fortunately been accompanied by thorough bridge deck waterproofing and so this and other concrete structures along the line to and from Mallaig should carry rail travellers for a considerable time to come.

9.3.2 Alkali–silica reaction

Some concrete bridge components have suffered from alkali–silica reaction. Almost all the confirmed cases, some 15, involve precast concrete parapet units originating from a British Rail works in Somerset. They were manufactured during the period 1969–1971 using a Portland cement which then contained just over 1% of soluble alkalis together with a fine aggregate now

Fig. 9.16 Frost attack, Glenfinnon Viaduct.

known to contain reactive silica. Some of the units have been replaced; others (Fig. 9.17) have remained under observation. Apart from some trials with protective coatings to keep out moisture, no repairs are contemplated.

9.3.3 Surface appearance

Bridge 733B, Eltham

Completed in 1987, this bridge was constructed to carry rail traffic over the Rochester relief road at Eltham. Civil engineers from British Rail's Southern

Fig. 9.17 Bridge parapet suffering fron alkali–silica reaction.

Region went to considerable lengths to ensure a satisfactory appearance. The box girders were to be cast *in situ* and the finish of the resultant 'highly visual' concrete would be a mixture of fair-faced and bush-hammered concrete. Placing difficulties demanded the use of very high workability concrete whilst strength requirements required a high cement content. Two admixtures were used to aid the exercise, whilst the pouring of trial panels would, it was hoped, reveal any shortcomings.

In the event the result was depressing. As Fig. 9.18 shows, wavy lines highlighted the top contour of the first pour, there were colour differences between bays and light and dark areas emphasized the boundaries between pours with the light zones being generally towards the bottom. None of this appeared during the trial pour. However, during the trial, unlike the pours

Fig. 9.18 Poor surface finish on bridge at Eltham.

themselves, there were no concrete delivery delays and no frustrating waits for placing to proceed as carefully planned.

Two main remedial options were considered: to bush hammer or grit blast the surface, or to apply a surface coating. The former would have the advantage of being cheaper and hopefully maintenance free, but would result in a reduction of cover and might not be totally successful. Coating, on the other hand, would improve the durability of the concrete and fully mask the flaws, but recoating would inevitably be necessary at some stage. Further, it would be more expensive than grit blasting. Because of a degree of uncertainty about the durability of coatings, it was decided to lightly grit blast the concrete and this has produced an acceptable finish.

9.4 The future

With each case of premature concrete deterioration, experience is gained which should prevent recurrences. At the same time, however, confidence in the use of concrete may be sapped. With so much concrete under its control, and more inevitably to come, it is not surprising that British Rail is now actively involved with durability research. With various collaborators British Rail has plain and coated reinforced concrete samples under test at various exposure sites. The project focuses on the prevention of deterioration of reinforced concrete structures in hostile environments and is being supported by Basic Research Industrial Technologies Europe (BRITE). A study of existing structures will supplement the exposure site data and both will provide the foundation of a life prediction model for concrete. Perhaps engineers will then specify maintenance-free concrete with confidence.

References

Child, C.A. (1988) Checking carbonation and chlorides on site, *Concrete*, **22** (5), 19, 20.
Dixon, J.F. and Sunley, V.K. (1983) Use of bond coats in concrete repair, *Concrete*, **17** (8), 34, 35.

Concrete pavements 10

G.D.S. Northcott
(Department of Transport)

10.1 Introduction

There are several different types of defect which occur in concrete roads. It is important that the type and cause is correctly diagnosed so that they can be effectively repaired. This chapter gives details of typical defects and the reasons for their occurrence, supported by examples in practice.

Until the early 1970s, concrete pavements in the UK were mainly jointed reinforced slabs. Jointed unreinforced slabs of the same thickness as reinforced slabs were then allowed as alternative designs. For the next 15 years concrete roads in the UK were constructed as unreinforced slabs which were more economical in initial cost. Current designs provide for thick jointed unreinforced slabs, thinner jointed reinforced slabs, continuously reinforced concrete pavements or thinner continuously reinforced concrete road bases with a bituminous surfacing. Figure 10.1 shows a typical longitudinal cross-section of each of these pavements. The contractor and engineer now have a wider choice of pavement to suit local conditions.

Concrete shrinks on hardening as discussed in Chapter 2, section 2.2.1. To allow for this and for the variations in dimensions which occur as a result of seasonal changes in temperature, joints or controlled cracks are necessary in pavement slabs. There are three types of joint: contraction, expansion and warping. Contraction joints allow slabs to shorten with a reduction in temperature. Expansion joints act similarly but also allow for additional expansion at temperatures higher than those at which the slab was constructed. Warping joints allow the slabs to curl when the temperatures of the upper and lower surfaces are different. As they are tied joints they only act as hinges.

Where there is a repetitive loading as in road slabs, there is a need for continuity between slabs to provide load transfer from one section to another to reduce rocking movement. In unreinforced concrete slabs this is provided by dowel bars or tie bars placed at mid-depth at the joint between slabs. Slab sizes are normally 5 m between joints but this can be increased to 6 m if limestone aggregate is used. In reinforced roads where slabs are longer the load transfer is provided by aggregate interlock at cracks which form between

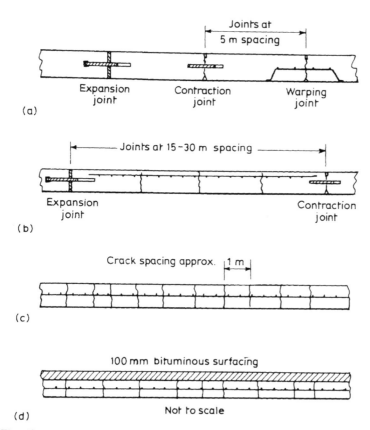

Fig. 10.1 Longtitudinal cross-sections of concrete pavements; (a) unreinforced concrete pavement; (b) jointed reinforced concrete pavement; (c) continuously reinforced concrete pavement; (d) continuously reinforced concrete roadbase.

joints and are maintained as fine cracks by the restraint of the reinforcement. The joints perform the same function as in an unreinforced slab.

Airfield pavements, where the loads may be greater but less frequent, are normally thicker unreinforced slabs, without dowels at joints, and laid on deeper sub-bases than road pavements.

10.2 Defects

10.2.1 Investigation

The Department of Transport's standard instructions (1985, 1988) for the inspection of trunk roads and motorways together with other local authority instructions give the required frequency of inspections, the methods of recording data and the reference points.

There is no easy way of assessing the number of defects in concrete roads without a close visual scrutiny. Video recorders can be used to picture the main defects, so that they can be viewed later, provided ambient conditions are such as to show up fine cracks, but the survey should be limited to one lane at a time. Cracks are best seen in cold weather when the surface is dry but before moisture in the concrete around the crack has dried. Computerized database systems are available which have been used to summarize the number of different types of defect in a section of road. However, they are not sufficiently advanced to be able to plot the data graphically in the form required by the Department of Transport for maintenance contracts. It will still be necessary to record the type, number and extent of the defects in fieldsheets on a grid related to traffic lanes and slab lengths. These are used as the basis for a schedule of different types of repairs for a maintenance contract.

From the initial and detailed surveys it will be necessary to carry out an assessment of maintenance required. For the safety of the travelling public emergency repairs may be required. These can be of a temporary or semi-permanent nature, and can be replaced later during a routine maintenance contract.

All other repairs, reinstatements or overlays should be properly programmed so that the various maintenance strategies and repairs can be evaluated to optimize the whole life costs.

10.2.2 Types of defect

The three main types of defect in concrete pavements — joint defects, surface defects and structural defects — are described in detail in *A Manual for the Maintenance and Repair of Concrete Roads* (Mildenhall and Northcott, 1986) together with recommended methods of repair. These are so written that they can be included in specifications for maintenance contracts. Summaries of these defects are given below for each of the three types.

Apart from normal wear and tear due to traffic loading and weather conditions, a number of defects have required early maintenance as a result of design, construction and maintenance faults, examples of which are given in later sections.

10.2.3 Joint defects

Shallow spalling

The largest number of defects at joints result from inadequate sealing

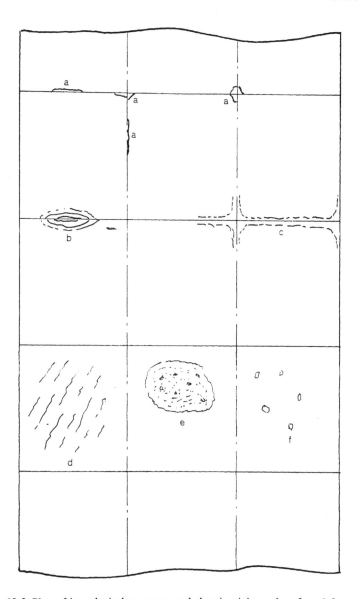

Fig. 10.2 Plan of hypothetical concrete road showing joint and surface defects; a, shallow spalling; b, deep spalling; c, cracking; d, plastic shrinkage cracking; e, surface scaling; f, pop-outs.

or poorly compacted low strength concrete near the joint. The seal debonds or tears from the slab allowing water and detritus to enter the joint.

In winter frozen water in the joint will cause spalling of the slab edges.

Without a proper seal, stones can enter and detritus can build up in the joint quite rapidly so that there is no expansion gap to cope with summer temperatures. The compressive forces at the bottom of the sealing groove will then cause spalling to the arrises (Fig. 10.2,a).

The risk of spalling is aggravated when the concrete around the joint former placed in wet concrete is not fully recompacted. Sawn joints do not have this disadvantage. Joint formers which are misplaced or tilted during construction will result in spalled arrises as any overhang will be broken off under traffic. This type of arris spalling can be repaired by thin bonded repairs.

Deep spalling

(Fig. 10.2, b) This is often misconstrued as arris spalling. However, the cracks, usually semi-circular in form, appear at a distance back from a joint of over 200 mm. It emanates usually from mid-depth as a result of dowel misplacement, locking of dowels or a build-up of detritus down in the joint crack. The build-up of detritus can be of such proportions that in high temperatures it will contribute to expansion failures (blow-ups) along the length of a traverse joint. These failures need full repairs.

D cracking

A series of small cracks adjacent and parallel to a transverse joint and continuing around the corner to a longitudinal joint is commonly called D cracking (Fig. 10.2, c). The cause is likely to be water-absorbent aggregates or porous concrete which are frost susceptible.

Misaligned crack inducer

Separate full depth cracks emanating from a joint to the outer edge or to a longitudinal joint are likely to be caused if the joint groove is not constructed correctly as a crack inducer. Without sufficient reduction of the slab section there is no particular place for the crack to form upon initial contraction and the crack appears off line. These require full depth repairs.

10.2.4 Surface defects

Irregularities in the form of bumps or low areas may be caused by faults at the time of construction. At a later period after trafficking, settlement of the subgrade and sub-base may result in cracking of the slabs and stepping between the slabs.

Plastic shrinkage cracking consists of a series of fine short parallel diagonal cracks in the surface not reaching to the edge of the slab and not more than 50 mm deep, caused by rapid drying of the concrete due to lack of proper curing (Fig. 10.2,d).

Surface scaling (Fig. 10.2,e) is a result of frost damage and a lack of entrained air. To prevent such wear, specifications for all road and airfield pavements require the concrete to be air entrained for at least the top 50 mm of the slab.

Loss of surface texture by rain damage during construction or by traffic later can be rectified by sawing closely spaced shallow transverse grooves in the surface. In order to prevent tyre noise the spacings have to be irregular, varying from 20 to 50 mm between grooves.

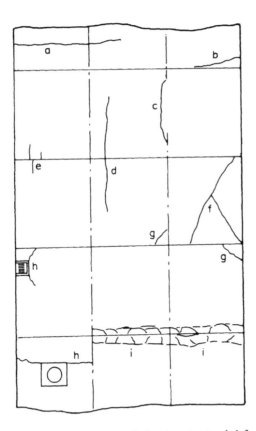

Fig. 10.3 Plan of hypothetical concrete road showing structural defects; a, transverse crack; b, crack at joint; c, longitudinal crack (joint former missing); d, longitudinal crack; e, short longitudinal crack; f, diagonal cracks; g, corner cracks; h, cracks at gulleys and manholes; i, expansion failure (blow-up).

Pop-outs are small defects in the surface where clay intrusions in the aggregate or other frost-susceptible pieces of aggregate cause local bursting of the concrete when frozen (Fig. 10.2, f).

10.2.5 Structural defects

Transverse cracking occurs at mid-bay or behind the dowel bars. It is likely to be due to the malfunction of adjacent joints which have not allowed sufficient contraction movement in winter, or to local settlement at mid-slab (Fig. 10.3, a).

Longitudinal cracking can result from omitted crack inducers (Fig. 10.3, c) or subsidence due to loss of sub-base support (Fig. 10.3, d). Short longitudinal cracks emanating from transverse joints can be shear cracks caused by higher compressive forces along one section of a transverse joint than the remainder (Fig. 10.3, e). This can be caused by restraint at one side of a slab, e.g. by a kerb or drainage channel tied to the slab without matching joints, or it results from a build-up of detritus in the low side of the crossfall that restricts movement on one side of the slab but not on the other.

Diagonal cracking and corner cracking (Fig. 10.3, f, and 10.3, g) are mostly associated with local subsidence under the slab. Small corner cracks at the intersection of the transverse and longitudinal joints arise from lack of compaction or an incomplete formation of the transverse joint.

Vertical movement at joints is due to poor load transfer at the joints, voids under the slabs, poor sub-bases and the ingress of moisture at the joint. This will lead to rocking of the slabs and mud-pumping at the joints.

Expansion failures (Fig. 10.3, i) can be dramatic blow-ups with one slab riding up on to the adjacent damaged slab. They occur in lengths subjected to a high compression in summer. The build-up of detritus in poorly sealed joints and an inadequate number of expansion joints contribute to the failure. After a number of seasons of expansion and contraction, the slabs creep up on the the sub-base and restrict the movement at certain joints.

10.2.6 Material defects

Concrete is affected by environmental conditions and by chemical attack. The most serious forms of damage that can occur to structural concrete are as follows:

1. chloride attack and reinforcement corrosion;
2. alkali–silica reaction;
3. carbonation;
4. sulphate attack.

These are considered below in relation to pavement concrete.

Chloride attack and reinforcement corrosion

Pavement concrete is generally made with a low water–cement ratio and is therefore much less permeable than some structural concrete. However, corrosion of reinforcement, dowels and tie bars does occur at joints or cracks in the concrete as a result of the use of de-icing salts. Unprotected dowel bars 25 mm thick have been known to lose half their diameter in 15 years and 12 mm tie bars in longitudinal joints have been completely eroded in a similar period.

The Department of Transport now requires tie bars to be protected against corrosion for a length spanning the joints. Dowel bars are usually sleeved to allow movement at one end. This sleeve has to be at least two-thirds the length of the bar to provide protection across the joint.

Mesh reinforcement in jointed reinforced roads has performed well without corrosion except where the surface or end cover has been inadequate, i.e. less than 35 mm. In continuously reinforced concrete (CRC) roads, where the cracks tend to be wider than in jointed roads, corrosion may be more of a problem where the steel is not at the centre of the slab. In CRC road design the reinforcement is now always placed at mid-depth. Epoxy-coated reinforcement could be used for CRC pavements for a longer life but is expensive. In the case of CRC road bases the bituminous surfacing provides considerable protection against de-icing salts.

Alkali–silica reaction

This has not been a problem in pavement concrete in the UK as there has been no evidence of map-cracking which is a symptom of alkali–silica reaction in structural concrete. Some traces of alkali–silica reaction gel have been found but only in isolated cases around cracks which have previously formed and allowed moisture to penetrate the concrete.

Carbonation

As carbonation occurs mainly in concrete which is kept relatively dry there is little evidence of carbonation in pavement concrete in the UK, where humidity is generally high.

Sulphate attack

Pavement design with its capping layers and sub-bases precludes attack on the concrete slab by sulphates from shales in the sub-grade. However, it is necessary to protect foundations and abutments to bridges by the use of sulphate-resisting cement or cement blends with pulverized fuel ash or ground granulated blastfurnace slag.

Design faults and remedies

10.3.1 Joint seals

Poor seals are often the main cause of defects at joints. In the early 1970s when unreinforced roads became economical the design rules of Road Note 29 (1970) allowed three out of every four contraction joints to be replaced by tied warping joints. This meant a distance of 20 m between moving contraction joints. However, as the slab lengths were only 5 m the joint groove size, which is dependent on the distance between joints, was still taken to be 10 mm instead of the 25 mm necessary for slabs 20 m long. This over-stretched the seals in the contraction joint and considerable spalling occurred at the contraction joints. In one road in the south east this affected nearly every contraction joint.

In two other roads the hot-poured type of sealant hardened with age after one summer and had no resilience the next winter. Another sealant was affected by moisture in the groove which caused bubbles and honeycombing to form, thereby reducing the bonding surface.

Problems also occurred in roads where a special type of compression seal was used which also acted as a joint former. The ends were sealed and the joint was evacuated to reduce its width before insertion into the fresh concrete. The ends of the seals were then cut to allow the seal to expand and to remain in compression at all times. On several occasions the seal remained the same thickness as when inserted and did not fill the gap at the joint (Fig. 10.4).

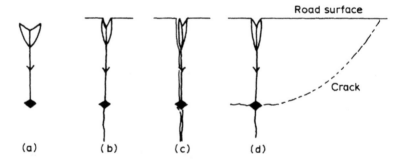

Fig. 10.4 Faulty compression seal in section: (a) tulip tree compression seal; (b) evacuated seal inserted into fresh concrete; (c) seal failing to expand to fill joint in winter; (d) diamond-shaped hard rubber knob causing horizontal cracking under compression and resultant spalling.

Fig. 10.5 Seepage of water over the hard shoulder
from beneath the concrete carriageway.

In one road where this type of seal was used it had a diamond-shaped hard rubber knob at the base of the fin (Fig. 10.4(d)). In later years, compressive forces around the knob caused horizontal cracks and spalling. The *Specification for Highway Works* (Department of Transport, 1986) now requires any protrusions on the fin of joint formers which are left in the concrete to be small and rounded in cross-section.

10.3.2 Deep spalling at joints

This has occurred as a result of the use of inadequate debonding paint which has caused locking of the dowels. Cracking at the reinforcement level has also resulted because of stress transmitted from the end of the dowels to the reinforcement.

10.3.3 Longitudinal cracking

For unreinforced roads designed to Road Note 29 bottom crack inducers could be omitted in summer on the basis that cracks due to warping stresses would start at the top. However, in the case of longtitudinal joints in two lane carriageways this joint may not form until much later in winter. At this time the crack emanates from the bottom and may miss the joint former at the surface. It is now mandatory to put a top and bottom crack inducer in the slab.

10.3.4 Cracks at manholes and gullies

Although manholes and gullies are separated from the main paving by expansion joint filler board this has not always provided sufficient allowance for the differential movement between the pavement and the manhole slab. Slab cracks emanate from the corners of the recess in the main slab (Fig. 10.3, h). To avoid this an extra joint should be included adjacent to the manhole.

Fig. 10.6 Slip-formed concrete pavement: a, poorly compacted permeable concrete at transverse joints in outer lanes; b, joint arris repairs to most transverse joints in centre and inner lanes.

10.3.5 Pumping

Bituminous hard shoulders were common with early concrete motorways, but their deeper construction led to water being trapped under the concrete carriageway and caused pumping at the longitudinal joint (Fig. 10.5).

10.4 Construction faults and remedies

10.4.1 Joint defects

Spalled arrises

If joints are formed in the wet concrete and the concrete is not fully

Figure 10.7 Spalled joint.

recompacted around the former, the void will fill with weak surface grout which will probably be porous and deteriorate under the action of frost and traffic. Most of the transverse joints shown in Fig. 10.6 have needed repair. In the outer lane of the near carriageway the dark concrete along the joint indicates wet porous concrete. Figure 10.7 shows the result of traffic wear on weak joints.

A survey by Gregory (1987) showed that this form of defect was more prevalent with slip-form pavers because of the method of recompaction over the former and the type of surface finisher used with these pavers. The method has been improved to ensure recompaction around and on top of the former.

Deep spalling

This can be caused by misaligned dowels which prevent movement.

Cracks at joints

Cracks emanating from joint positions which wander towards the outside edge of the slab can occur when joint grooves are sawn or formed.

In sawing joints it is necessary to saw as early as possible after the concrete has hardened without causing spalling by plucking out of the coarse aggregate. If left too late, the concrete may already be in tension and the cutting of the groove from one side will initiate a crack which, without a discontinuity to follow, will take the easiest route towards the edge (Fig. 10.3, b).

Wide dual carriageways with several traffic lanes are often constructed in two halves longitudinally as the width of the slabs is dependent on the maximum width of the paving train. If laid in summer with wet-formed joints only one joint in about every four will crack initially due to shrinkage of the concrete (Fig. 10.8, joint type A). These joints tend to have greater movement until such time as the remaining joints crack (Fig. 10.8, joint type B).

If the second half (or rip) is laid adjacent to the first, the pattern of initial cracking may not coincide with that of the first rip. When winter comes the joints in the first rip will open up and induce tensile stress at the side of the adjacent second rip and cause the joint to crack. The crack will emanate from the butt joint towards the remote edge. If there is a proper discontinuity along the line of the joint which has been formed correctly with a joint groove then the crack will follow the discontinuity.

If the joint former is buried or partially omitted so that there is no major discontinuity for the crack to follow, it will be deflected off line by the coarse aggregate or by following microcracks within the concrete matrix (Fig. 10.8, joint type C).

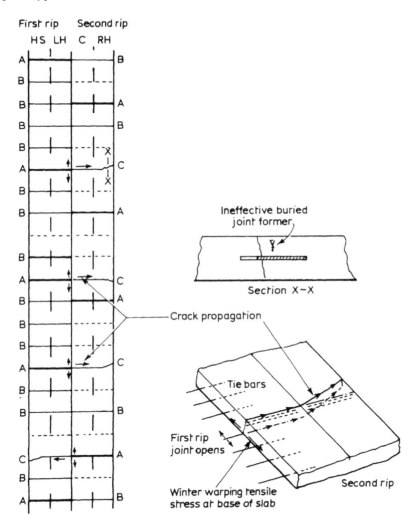

Fig. 10.8 Sympathetic cracking at joints: A, joints cracked as a result of initial shrinkage of concrete on hardening; these joints tend to have greater movement; B, additional joints cracking later, due to seasonal reduction in temperature or diurnal warping, together with construction traffic loading; C, cracks forming at joint positions in second rip as a result of the influence of movement or joints in the first rip without a proper discontinuity or notch in the top of the slab.

10.4.2 Surface defects

Surface irregularities

These can result from grout runs at the edge of badly adjusted brush-texturing machines. Bumps can also occur when the paving machine is stopped. Isolated cases will not impair the riding quality unduly, but if the stoppage is regular, as in one road where joints were formed by stopping a slip-form paver, the bumps become cyclic and create an unacceptable ride at certain speeds. Bumps can be rectified by bump cutting using grinding apparatus or carefully controlled bush hammering within contours marked out after using a straight edge and wedge.

Stepping between slabs indicates a failure of the dowel bars associated with subgrade failures. To rectify this the voids beneath the slab can be filled by vacuum grouting, but failing this the slabs should be replaced.

Surface scaling

Frost damage at the surface is usually the main cause of widespread surface scaling and is brought about by lack of or insufficient air-entraining agent. It has occurred in very workable mixes with high water–cement ratios which, with low air content, produced permeable concrete at the surface in hand-laid sections of slip roads.

The following remedies could be used depending on the severity of the damage.

1. First carry out any necessary bump cutting; then reduce the permeability of the surface concrete by impregnation using a silane-based material. This is likely to reduce the skid resistance of the concrete slightly.
2. Remove the top surface of the slab by bush-hammering 30–50 mm down and replace by a thin bonded concrete topping.
3. Break out and replace the whole slab.

Plastic shrinkage cracking

This shows up as a series of fine parallel short diagonal cracks in the surface and in the centre of the slab. These are caused by lack of suitable curing especially with concrete of a high water–cement ratio (Fig. 10.2, d).

If the cracks are fine, impregnation using silane will reduce the possibility of ingress of water and frost damage. For it to be effective, the impregnation should be done when the concrete is dry. Alternatively, the top 50 mm approximately should be removed and a thin bonded concrete topping applied.

Pop-outs

Small round holes appear in the surface of the concrete after one winter (Fig. 10.2, f). These are caused by frost-susceptible material in the aggregates. In some cases it has been caused by clay lumps from the base of the aggregate stockpiles and to avoid this bases of stockpiles should be cement treated. In another case chalk lumps were mixed with the flint coarse aggregate and caused a large number of pop-outs in one section of motorway.

Some flints from Hampshire and Sussex which appear whiter than normal tend to be particularly frost susceptible. This is because the white cortex material surrounding the flint has a high water absorption. Pop-outs have occurred using this type of aggregate on airfields where the debris is much more of a safety hazard than on roads where it is simply a nuisance. If there is sufficient of this aggregate in the mix it can expand in winter and cause D cracking (Fig. 10.2, c). The aggregate can also contribute to alkali–silica reaction.

Small pop-out holes can be reamed out and plugged with a 1:3 cement:sand mortar. D cracking down to mid-depth is much more costly to repair as it requires full depth repairs.

10.4.3 Structural defects

Transverse cracks

These occur at mid-bay or behind the dowel bars and are often caused by locked joints, misplaced dowel bars or irregularities in the sub-base. In one case the bitumastic paint used as a debonding agent for the dowel bars was discovered to be a roofing felt adhesive. As a result most of the joints were tied and cracks appeared at mid-bay.

The remedy is to carry out a full depth repair and form a new joint near the crack or undertake complete bay replacement, depending on the degree of damage.

Fig. 10.9 Stitched crack.

Longitudinal cracks

These appear in slabs when the longitudinal joint formers and crack inducers have been omitted.

If the crack is within the middle third of the tie bar it need only be routed and sealed. If outside this limit then the crack should be stitched by inserting from the surface additional staple-type tie bars. These are embedded in epoxy resin mortar within a slot cut in the concrete which is subsequently filled with fine concrete or cement–sand mortar (Fig. 10.9).

Diagonal cracks

These are symptomatic of sub-base and subgrade settlement. The most common place for them to appear soon after construction is next to the underbridges where the backfill to the bridge has not been fully compacted.

In such a case the cracks will quickly widen. In this situation full bay replacement should be considered after further compaction of the sub-base. Elsewhere, fine diagonal cracks can be contained by stitching with staple tie bars, but if the cracks are wide and accompanied by considerable settlement the bay should be replaced.

Fig. 10.10 Overband sealing of joints.

10.5 Maintenance faults and remedies

10.5.1 Joint sealing

Joints have been resealed by overbanding them with a bitumastic sealer. Apart from the poor aesthetic appearance (Fig. 10.10) overbanding tends to be a temporary seal as it will crack with the movement of the joint (Fig. 10.11). It also disguises the degree of arris spalling which makes surveying for a repair

Fig. 10.11 Ineffectiveness of overbanding

contract difficult. An efficient way of removal of overbanding sealing is by high pressure water jetting.

10.5.2 Joint arris repairs

Thin bonded repairs have failed (Fig. 10.12), sometimes after a relatively short period, for the following reasons:

1. wrong diagnosis of the original fault;
2. lack of curing
3. use of wrong materials;
4. lack of compaction;
5. patching over a joint;
6. inadequate preparation of the substrate.

The more deep-seated spalling has often been treated by a thin bonded arris repair, and after a period of trafficking the cracks in the parent concrete will reflect through the patch. These need to be replaced by full depth repairs.

As a result of lack of curing in hot weather, patch repairs, which tend to shrink inwards on setting, will warp and reduce the bond at the edges. The use of curing compound alone is insufficient with such patches. They should also be kept damp and protected from the sun and wind for as long as possible while the road is closed.

Fig. 10.12 Failed arris repair

To reduce road closure time, fast-setting materials such as high alumina cement mortars have been used. However, in wide patches the heat of hydration of this type of mortar can lead to warping of the patch and lack of bond at the edges.

Lack of compaction around the edges of the patch allows moisture to enter and frost damage will occur.

If repairs are required at both sides of a joint it is necessary to provide a flexible groove former over the previous joint between the patches. Even if grooves were later to be sawn for sealing it would be difficult to follow the resultant crack in a patch which straddled the joint.

Failures of patches have occurred as a result of the main slab being too dry before concrete is applied. To achieve a good bond with cementitious materials it is necessary to soak the slab for several hours prior to applying mortar or fine concrete. However, in the case of epoxy resin repair mortars the concrete should be dry.

10.6 Repair materials

There are a number of different proprietary brands of repair materials (Chapter 4, section 4.2); however, experience has shown that cementitious materials are more effective with concrete slabs.

For temporary emergency repairs in the interest of vehicle safety, bituminous materials, thermoplastic materials or epoxy resin mortars can be used in potholes caused by spalling or cracking.

If time allows for proper preparation of the holes to be patched these materials will have a longer life. Their advantage is that the road can be opened to traffic early but the disadvantage is the poor aesthetic appearance of black patches in the concrete.

More permanent rapid repairs can be achieved by the use of the following:

1. high alumina cement;
2. magnesium phosphate cement;
3. rapid hardening cement;
4. superplasticized concrete;
5. water reducers and workability aids to achieve early strength by reducing the water–cement ratio.

With the use of these rapid-hardening materials, the time before opening to traffic can be reduced considerably. In the case of rapid-hardening cement or magnesium phosphate this can be about six to eight hours. These quick-setting materials are difficult to work with and should be restricted to small patches and for arris repair rather than for full depth repairs.

Cementitious mortar and concrete should be used at other times in maintenance contracts; road closures are required for the preparation of repairs and should be maintained for the full curing period required for concrete strength development. Curing times can be shortened to the time when the compressive strength of the patch material reaches $25\,\text{N/mm}^2$ rather than waiting for the full period normally required for new work (i.e. to achieve a strength development of $40\,\text{N/mm}^2$).

10.7 Overlaying and strengthening

Although pavements are designed for a finite number of years on the basis of traffic forecasts, the actual period may be shorter if traffic and deterioration has increased. However, with regular maintenance and slab replacement it could be considerably increased. Overlays, which in the UK tended to be bituminous materials, need to be applied to the whole of the carriageway irrespective of whether the whole carriageway needs repairing.

Before embarking on an overlay there is a need to consider the overall long-term costs of the different maintenance options:

1. piecemeal repairs;
2. local reconstruction, e.g. sections of one lane;
3. total overlay.

These costs should include those of traffic delay for the road closure time in addition to the cost of the repairs themselves. The whole life costs of the various options should be compared, taking into account the frequency of further repairs and the need for the replacement of overlays.

On very heavily trafficked roads such as the M1, frequent closures for minor repairs are unsuitable. For roads of this type major maintenance contracts at intervals of over five years are appropriate. Older concrete roads with lower traffic densities can be maintained successfully by annual repair contracts phased into a five year programme.

References

Department of Transport (1985) *Code of Practice for Routine Maintenance*.
Department of Transport (1986) *Specification for Highway Works*, HMSO, London.
Department of Transport (1988) *Visual Condition Surveys of Concrete Pavements*, Departmental Standard HD17/88, Department of Transport, London.
Gregory, J.M. (1987) *The Performance of Unreinforced Concrete Roads Constructed between 1970 and 1979*, TRRL Research Report 79, Crowthorne.
Mildenhall, H.S. and Northcott, G.D.S. (1986) *A Manual for the Maintenance and Repair of Concrete Roads*, Department of Transport, Cement and Concrete Association, HMSO, London.
Road Note 29 (1970) *A Guide to the Structural Design of Pavements for New Roads*, Department of Energy – Road Research Laboratory, HMSO, London.

Marine structures   11

M.B. Leeming
(Formerly Arup Research & Development)

11.1 Introduction

Four environmental zones are usually identified on both coastal and offshore structures: the underwater zone, the tidal zone, the splash zone and the atmospheric zone. Each has different requirements for repair, different forms of damage and deterioration and different problems associated with it. The tidal and splash zones are closest in nature and have many similarities. For coastal structures, particularly in harbours and estuaries, the regular rise and fall of the tide is the dominating factor, while for offshore structures the tidal range is generally small and the main factor is wave action. The atmospheric zone has few special problems compared with a land structure except perhaps access and a dominance of chloride-related deterioration. These aspects have either been dealt with elsewhere or are less severe cases of what is required in the splash and tidal zones and will therefore not be dealt with specifically here. Consideration of repairs to marine structures thus concern two main areas: the underwater zone and at water level and above.

Offshore structures for the production of oil and gas are relatively new as a breed of marine structures, the first structure, Ekofisk, being installed in 1973. In the last 15 years these structures, which were constructed to high standards, have given good service and have not suffered any significant deterioration. Considerable attention, however, has been paid to the problem of their repair, not so much because of the necessity of repair, but because of the logistical problem of carrying out repairs in deep water at a distance from land and because of the considerable loss of revenue should it be necessary to take them out of service.

All the considerations with respect to the durability of materials and the methods of making repairs to any marine structure apply equally to offshore structures. However, there are a number of additional considerations that are particular to offshore structures, notably the need to effect the repairs in a much more critical weather window and the problems of making repairs in a considerable depth of water.

Apart from cosmetic reasons there are two distinct reasons why a repair needs to be carried out on a marine structure. Repairs either can be necessary to maintain structural integrity or are needed to arrest deterioration of the concrete, particularly with regard to corrosion of the reinforcement or the prestressing tendons.

11.2 The need for repairs

11.2.1 The underwater zone

Recent research (Leeming, 1989) into the corrosion processes in marine concrete has shown that, under water, corrosion of the reinforcement does not occur in normal circumstances. However, where significant areas of reinforcement are directly exposed to seawater, e.g. in cracks over 0.6 mm wide or in areas where impact damage has occurred, corrosion can occur but this can be controlled by the application of cathodic protection. Most offshore structures have to date been cathodically protected to protect steelwork attached to the main structure such as riser pipes. As it is difficult to ensure that the steel reinforcement is not electrically bonded to any fixings to the attached steelwork, the cathodic protection net is usually extended to take in the reinforcement embedded in the concrete. The above areas may need repair if cathodic protection is not applied, or if it is desirable to reduce the drain on the cathodic protection system.

Concrete under water may suffer accidental damage or foundation failure or may be found to be structurally inadequate and therefore repair may be required to maintain structural integrity. Coastal structures sometimes suffer loss of concrete section at sea-bed level due to abrasion from water-borne sediments and repair may be required to maintain serviceability.

11.2.2 Above water

The area of the structure just above the low water level, usually called the splash zone offshore or the tidal zone at the coast, is the most vulnerable with regard to deterioration of the concrete due to corrosion of the reinforcement. This leads to spalling of the concrete and rust staining long before loss of section of the reinforcement becomes structurally significant. However, repair is required as the rate of corrosion increases once the cover of concrete to the reinforcement is lost.

This area is also vulnerable to physical damage due to shipping impact and to abrasion from fenders, mooring ropes and wave action. Conventional cathodic protection with impressed current or sacrificial anodes immersed in the water is effective only for areas under water. However, techniques incorporating anodes formed from arrays of wires or conductive coatings have been developed for use on bridge decks, piers and abutments (Chapter 5, section 5.4). These methods are still regarded by many as in the development stage although they are now being applied experimentally to coastal structures.

11.2.3 The repairing of cracks

The main reasons for repairing cracks are as follows:

1. to stop leaks;
2. to maintain structural integrity across the crack;
3. to prevent ingress of aggressive species that cause corrosion of the reinforcement;
4. for cosmetic reasons.

A thorough analysis is necessary to determine the cause or causes of any cracks and to decide on the necessity for their repair.

11.2.4 The durability of repairs

It is the proximity to the sea which provides a harsh environment for reinforced concrete resulting in a greater need for repairs. The repairs have to withstand the same harsh environment. Additionally, repairs in a marine situation are more difficult to carry out, placing a greater premium on good workmanship. In many instances it is not possible to keep the repair area dry and clean which makes it more difficult to achieve a sound repair. The high humidity and ingress of chlorides can affect the structural performance of the repair and the ability of the repair to control further corrosion damage. Special methods and materials are required to overcome these problems. The problems of access and the difficulty of getting the work done in marine conditions make the choice of the right method and material much more important from an economic standpoint.

There are four main aspects that can affect the durability of repair systems in marine situations.

Material degradation

The majority of materials used for the repair of reinforced concrete have qualities, with regard to durability in a marine environment, which are as good as, if not better than, those of the parent concrete. This statement, however, must be made with some caution as resins and polymers are relatively modern materials which have not withstood the test of time as long as reinforced concrete has. There have been instances where materials have deteriorated rapidly (John *et al.*, 1983), but these have been caused mainly by wrong formulation, inaccurate proportioning, inadequate mixing and poor application. These conditions generally result in a very porous material and the deterioration process is due to the action of water. Quality assurance, proper supervision and site testing should avoid these problems. Past experience is the most reliable source of information on the problems and durability generally, but historical information in the form of case studies is not easy to come by.

Poly vinyl acetate was found not to be resistant to saponification in permanently wet conditions and is therefore no longer used in marine conditions. There have been reported cases of what were thought to be sulphate attack both to styrene-butadiene rubber modified cement mortar containing glass fibres and to a rapid-setting repair system (Coote *et al.*, 1986). Other possible causes of material degradation are the action of sulphate-reducing bacteria, the ingress of significant quantities of chlorides and high hydrostatic pressure due to deep immersion. These problems have been little researched but it is believed that present polymer systems, when correctly formulated and applied, are at least as durable as well-constructed concrete and they have generally behaved well. It is therefore thought that material degradation even in the long term is not a major problem.

Interface bond

Poor interface bond could cause a lack of durability in marine situations, leading to poor structural performance and detachment of the repair material, and it can provide a path for aggressive agents that cause corrosion of embedded reinforcement. It is a problem that can occur with all repair materials under all conditions, but marine conditions can present special problems such as contamination of the surface of the concrete to be repaired with chlorides and with algae and marine growth. These problems relate to the greater difficulty in surface preparation and to the contamination of the surface with chlorides particularly when the repair is carried out under water.

Corrosion protection of the reinforcement

Continued corrosion of embedded reinforcement will cause further deterioration of the repaired structure and will possibly disrupt the repair material. Often the objective of carrying out a repair is to stabilize or minimize further corrosion whether the original damage was caused by impact or by spalling due to corrosion. In instances where the cover to the reinforcement was initially low, the replaced repair materials may need to perform better with regard to corrosion protection than the original concrete. The mechanisms of corrosion protection provided by concrete to the reinforcement in a marine environment are not fully understood although recent research has considerably advanced present knowledge.

Other than the work by the Building Research Establishment referred to above (Coote *et al.*, 1986), the mechanisms of corrosion protection to the reinforcement provided by various repair materials have not been studied in detail. There is a fundamental difference in the mechanisms of protection provided by mainly cementitious and purely resin repair systems. The former attempts to replace the contaminated or damaged concrete with a material similar to the original, hopefully returning the reinforcement to an alkaline environment and thereby repassivating it. Resin repair systems provide a purely passive protection by attempting to encapsulate the reinforcement in an inert material impervious to corrosive influences.

The insertion of repaired patches into a structure further complicates the understanding of the mechanisms of corrosion in concrete. There is evidence to show that freshly repaired areas can accelerate the corrosion of adjacent areas of chloride-contaminated concrete (Cavalier and Vassie, 1981). Back-diffusion of chloride ions from the parent concrete into the repair material may be a factor that could influence the durability of repair systems.

Physical Factors

Resistance to surface scour (abrasion), impact damage, fire and freeze–thaw cycles should be considered as influences on the durability of repairs. However, there is little information as to the performance of repair materials with respect to these factors.

11.3 Repairs to deteriorated concrete

11.3.1 Spalling and reinforcement corrosion

The repair of concrete damage arising from reinforcement corrosion involves more than just replacing the concrete that has spalled. It is also necessary to replace all the concrete around the reinforcement that contains a sufficient quantity of chlorides to cause reinforcement corrosion in the future. This requires a detailed survey of the extent of chloride ingress. It is then necessary to make a decision as to the percentage of chloride ions by weight of cement that is likely to cause corrosion. This is no mean task as the experts disagree on this value. The Building Research Establishment (1982) have said that 0.4% of chloride ion by weight of cement will give a medium risk of corrosion. Other work by the Federal Highways Administration in the USA (FHA, 1987) has shown that a lower value of 0.2% is the corrosion threshold level for moist cured concrete. The difference between these values can make a considerable difference to the overall extent and cost of the repairs required.

It is then necessary to decide on the thickness of concrete contaminated to a lesser degree that can be allowed to remain around the bar to allow for future chloride diffusion or redistribution. This thickness depends also on the measures taken to prevent or slow down further chloride ingress. Such measures may include the use of surface coatings on the concrete which have the benefit not only of providing protection but also of covering up the repairs (Chapter 5). However, marine conditions are very severe on any thin surface coating and there can be the added problem of saline water rising behind the coating due to capillary action and causing further chloride ingress and damage to the coating. All the contaminated concrete must be cut out and replaced with a repair system which is impervious to further chloride ingress through the surface and from the surrounding contaminated concrete.

11.3.2 Repairing cracks

It is first necessary to determine whether the crack is still moving as this determines the method and material used in the repair. If no further movement is taking place the crack can be grouted up with cement grout for wide cracks (above 1.5–2 mm) or resin for narrower cracks as it will give greater penetration. Small movements in narrow cracks can be dealt with by the use of polyester resin which has a small amount of flexibility. If large movements are experienced which cannot be halted it is necessary to rake out or form a groove along the crack to take some form of joint filler or sealing compound of sufficient dimensions to accommodate the movement (Chapter 4, section 4.2.6). Few methods of joint sealing by grouting or otherwise can provide much tensile strength across the crack but compressive forces can easily be accommodated. If tensile forces are to be taken across the crack without further cracking then it is necessary to provide additional reinforcement or prestressing to take the tensile forces.

11.3.3 Repair of construction faults

During the construction of many of the concrete offshore oil platforms, repairs were carried out on the faults that arose through bad workmanship and defective materials. Little of this work has been described in the literature. It is a very worthwhile strategy as the extent of the repairs is considerably less than would soon result on exposure to the sea and they are much easier to carry out. The repairs amount to rectifying areas of low cover and cutting out and replacing honeycombed concrete and other areas of high porosity. In many instances this work can be carried out by recasting in normal concrete provided that adequate provision is made to bond the new work to the old and to avoid shrinkage cracks

11.4 Repairs necessitated by damage to offshore structures

11.4.1 Shipping damage

Shipping damage is a major concern for offshore structures and there have been several studies of the likelihood of shipping impact and the likely consequences (Department of Energy, 1984; Wimpey Laboratories, 1987a). While the collision of a large ship with an offshore structure has yet to happen there have been several collisions of supply ships with steel and concrete structures. Shipping impact with concrete structures in harbours and estuaries is a regular occurrence. The most likely form of damage is shear and bending failure of beams and piles and punching shear to the hollow legs of offshore structures. Except in the most serious of collisions the damage to an offshore structure is not thought to be critical to the collapse of the structure. Nevertheless, leakage of water into damaged towers could have serious consequences for production, and in addition many structures rely on external water pressure to apply hoop prestress to the tower. Serious damage from shipping impact generally involves reconstruction of the member.

11.4.2 Damage due to dropped objects

Damage due to dropped objects is particular to offshore structures. When loading material onto the platform, drill strings, pipes and other bulky objects can drop during lifting and damage or puncture the roofs of oil storage cells. This damage could in certain cases lead to the spillage of oil into the sea. Knowing the weight and shape of the object and the height of the fall it is possible to assess the likely damage (Wimpey Laboratories, 1987b). The degrees of damage could be penetration and front-face spalling, penetration and spalling together with scabbing of the back surface, shear plug formation and, ultimately, complete perforation. The damage needs to be repaired to stop leakage either into or out of the cell. Generally at the depths at which the cell tops are located corrosion of the reinforcement is not a problem. Because of the depth, however, the damage is difficult to inspect and to repair. Repair schemes have been worked out in advance by most oil companies so that the operation can be put into immediate effect should this be required.

11.4.3 Fire damage

Fire is another hazard for offshore structures and studies have been instituted to assess the likely damage from possible scenarios (Marine Technology Support Unit, 1986). Jetties such as Bantry Bay have also been damaged by fire. The damage from small fires results generally in spalling of the concrete surface. It becomes necessary to hack off all the loose concrete and to replace the cover to the reinforcement to avoid subsequent corrosion damage. Larger fires may cause much more serious structural damage and cracking of members. Repair in these situations becomes a major rebuild.

11.4.4 Gouges and abrasion

Superficial damage is often caused by the action of mooring ropes, chains and other metal objects rubbing against the side of the concrete. Abrasion of concrete surfaces can also occur due to sediments being dashed against the concrete surface by wave action. This damage generally reduces the cover of concrete to reinforcement and needs to be repaired to prevent corrosion of the reinforcement.

11.5 Case studies of repair methods

Repair methods in marine situations are difficult to categorize as there are many constraints and each situation requires its own solution. A study of methods is best approached from particular case studies presenting the solution adopted. A summary of the repairs that have been carried out to existing off-shore structures is given in Table 11.1.

11.5.1 Shipping damage to the leg of an offshore structure

A supply boat hit the leg of a 400 mm thick reinforced concrete caisson wall of an offshore installation (Faulds *et al.*, 1983; UEG, 1983). A local punching shear failure resulted, with associated cracking over an area some 4 m high and 4 m wide around mean sea level. Water flowed intermittently through the cracks. The cracks were first sealed using a combination of epoxy to seal the cracks and gunite concrete applied to the outside of the wall. Once the flow of water had ceased the inner half of the wall was broken out over the damaged area and replaced by new concrete cast *in situ*.

In the following summer season a cofferdam was fitted around the damaged area on the outside of the leg and the outer half of the wall was broken out in the dry. Concrete was then cast in the repair area. The cofferdam gave a working space 1.5 m deep. Resin injection points were cast into the new section so that any area of imperfect bond between the old material and the new could be grouted up. There were difficulties in fixing the cofferdam in the exposed conditions of the open water in the middle of the North Sea.

11.5.2 Punching shear failure to a caisson roof

A length of a 36 inch diameter pipe weighing 10 tonnes fell on to the 500 mm

Table 11.1 Existing concrete structures and repairs carried out

Concrete platform	Year of installation	Operator	Cause of damage	Damage caused	Extent	Reason for repair	Depth of repair below sea level (m)	Material used	Method	By whom carried out	References in UEG Publication UR 36	Cost (£ million)
Ekofisk	1973	Phillips		No information								
Beryl A	1975	Mobil	Construction	Cracking during construction	–	–	–	Cement grout	Injection	–	–	
Brent B	1975	Shell	Riser pipe	Gouged shallow groove in leg	Superficial	Restore cover	35	Epoxy	Fill shutter	McAlpine Sea Svcs	20,23,24	0.4
			Fire damage	Spalled cover	Superficial	Restore cover	Air	–	–	Yet to be carried out	23	
			36 inch riser	Hole in caisson roof	Major	Reinstate	90	Concrete	Shutter, place agg, grout	McAlpine Sea Svcs	19*, 20,22 23,28* 29,16	1.4
Frigg CDP1	1975	Elf	Foundation failure	Cracking to diaphragm walls	Major	Reinstate	80–90	Epoxy	Injection	Elf	21*,23	
Frigg TP1	1976	Elf	Built-in steelwork	Leak	Minor	Stem leak	–	Cement grout	Injection	Elf		
Frigg MCPO1	1976	Total		No major repairs							–	
Brent D	1976	Shell		No repairs								

Stratfjord A	1977	Mobil	Fire damage	—	Superficial	Restore cover	—	—	—	—	23	
			Submergence	Cracks leaking into cells	Major	Stem leak	110	Epoxy	Injection	Norwegian contractors	26,27*	
Dunlin A	1977	Shell	Drain caisson	Chipped groove	Superficial	Restore cover	120	Epoxy	Fill shutter	McAlpine Sea Svcs	20,22,23 24,16	1.5
Frigg TCP2	1977	Elf		No information								
Ninian Central	1978	Chevron		No repairs								
Cormorant A	1978	Shell		No repairs								
Brent C	1978	Shell	Ship impact	Through cracking to shaft	Major	Reinstate	0–2	Concrete	Form in dry	McAlpine Sea Svcs	20,22,23 24,28*,29 16	1.0
Stratfjord B	1981	Mobil	Construction	Lack of cover	Superficial	Restore cover		Epoxy	Coating	Norwegian contractors	26	
Stratfjord C	1984	Mobil		No information								

thick reinforced concrete roof to a cell of a concrete gravity platform. The roof was at a depth of 80 m. The pipe cut a neat 1 m diameter hole more than half way through the slab with associated cracking forming a shear plug. As the cells were maintained at an internal pressure of 4 bar below the external seawater pressure there was an immediate flow of water into the cell. As the under-pressure was designed to provide a compressive prestressing force to the concrete it was essential to stem the flow of water. An immediate temporary repair was effected by placing a steel cover plate over the affected area and sealing it with sandbags.

A permanent repair scheme was then planned. The steel plate was replaced by a new 50 mm thick steel plate with epoxy resin seals and the void below was grouted with epoxy resin. This repair could withstand the normal operating pressures but not the maximum required design pressure. Steel tie bolts were drilled into the existing concrete dome and a shutter was placed around the repair area. The shutter was prepacked with reinforcement and coarse aggregate and then closed with a lid. A sand–cement grout was then injected into the shutter from a surface support vessel. The method of repair was first tested on shore.

11.5.3 Gouges in concrete cell tops and caisson walls

The first gouge was caused when a 30 inch diameter drain casing broke loose from its support and fell end-on onto the caisson roof below at a depth of 120 m. The damage was superficial and no reinforcement was exposed nor was there any cracking. The objective of the repair was to reinstate the cover to the reinforcement. The damaged area was filled with epoxy resin by injecting behind a shutter fixed to the concrete. Special injection equipment was devised to ensure safety for the diver. The high bond strengths of the specially developed resins for underwater use are excellent for small superficial repairs but for large volumes the exothermic reaction becomes a major problem. Their low modulus of elasticity also makes them unsuitable for repairs where structural strength is required.

The second gouge occurred when pipework broke loose causing a gouge in the concrete leg 35 m below sea level. Damage was again superficial but two reinforcing bars were exposed. Corrosion protection was again restored in a similar manner to that described above.

11.5.4 Damage to piles, jetties and other coastal structures

A large number of repairs have been carried out in a marine environment to jetties, docks, bridge piers and concrete ships but the details of only a small number have been published. A UEG report (UEG 1986) on the durability of repairs to both coastal and offshore structures reviews the evidence presented at conferences on the maintenance of maritime structures (ICE, 1977; 1986), on the performance of concrete in a marine environment (ACI, 1980) and on the repair of reinforced concrete in the Arabian Gulf (CIRIA, 1985).

Six case studies of the deterioration and repair of coastal structures in the UK, including jetties, piers and quays, discuss the reasons for failure and deterioration. It is concluded that chloride-induced corrosion of the reinforcement is the major cause of deterioration and occurs mainly between the upper limit of marine growth and the top of the splash zone, deck support beams being the worst affected. Impact and abrasion are also significant contributory factors.

In an overview on the repair of typical coastal structures, Glassgold (1982) briefly describes deterioration characteristics. Three basic zones are analysed for repair procedures (submerged, tidal and exposed) although there are essentially only two basic approaches for repairing structures in seawater: 'dry' and 'wet'. 'Wet' applies to repairs where the water is displaced by the repair material while 'dry' assumes some sort of cofferdam and dewatering. Workmanship and quality of materials need to be higher for the 'wet' process while working in the 'dry' allows evaluation of the problem, greater flexibility in the choice of material and method and better inspection, supervision and control. 'Dry' repairs are the desirable choice except when they are physically impractical or uneconomic. These remarks mainly apply to submerged repairs but also to those in the tidal range.

The above conferences also provide a valuable series of case histories of repairs carried out to coastal structures throughout the rest of the world and describe methods and materials used and causes of deterioration. Several of the papers describe the results of short-term evaluation trials but none of them gives any information on the long-term durability of repair systems. Shotcrete is extensively used in the USA as a repair method but experience from periodic in-service inspection of concrete offshore structures has shown that shotcrete used to compensate for minor deficiencies in the cover thickness in the construction phase of these structures has a tendency to spall off in service.

11.6 Repair materials

11.6.1 Repair systems

Repair methods and repair systems are described in detail in Chapter 4 and also in the Concrete Society *Technical Report 26* (Concrete Society, 1984). Marine applications make special demands on the various parts of the repair systems and the material used. Only repair systems which the supplier has demonstrated are satisfactory for marine use should be specified.

11.6.2 Testing repair materials

The economics and difficulties of carrying out repairs in a marine environment provide a strong argument for thoroughly testing repair materials and carrying out research into the durability of repair systems. However, there are equal difficulties in doing research in the field. These are set out in a UEG report (UEG 1986). Carrying out repairs is a labour-intensive operation with the major cost arising from access to the site and preparation for the repairs, e.g. breaking back to sound concrete and fixing of formwork, which are common to all systems. The final preparation and the choice of the right material can have a disproportionate effect on the durability, and therefore final economy, of the complete operation.

At present there is little information on the durability of repair systems both onshore and for marine applications. The choice of material is made largely by guesswork and by experience of what has been adequate in the past. This is not conducive to improvement. Many structures are now being repaired at an age of less than 20 years. Few repair systems have demonstrated reliability for periods in excess of this. These facts suggest that continued maintenance must be accepted for structures that were designed for lives well in excess of 20 years.

Research on the durability of repair systems will give confidence in the choice of materials and will result in repairs being carried out more effectively. It will allow improvement in repair materials and systems as knowledge is gained of the mechanisms involved and as the important parameters affecting durability are identified.

Relevant codes and standards

There are few standard tests for repair materials, although there are moves to redress this within the British Standards Institution and the associated European Standards Committees. These tests, when adopted by the appropriate authorities, will probably be directed towards repairs to materials for onshore buildings and bridge structures. The more severe marine environment will demand more rigorous tests.

Material degradation

The particular aspects of material degradation of repairs to marine structures were discussed in section 11.2.4 above. Clearly the materials need to withstand long-term saturation in seawater. The supplier should produce evidence that the materials are satisfactory for a marine environment.

Interface bond

As discussed in section 11.2.4 the junction between the repair material and the parent concrete is a critical area for durability. A firm bond is required which needs to be made under the onerous conditions of damp concrete and possible contamination by salts, silt and marine organisms. A tolerant and efficient bonding bridge is required. The specimen preparation for any interface bond test should reflect the conditions found in a marine situation.

Corrosion protection to the reinforcement

The requirements for a repair material to provide protection against further corrosion of the reinforcement in a marine situation are the same as those for onshore bridge structures. The environment, however, is more severe. The repair material needs to provide the protection under more saturated conditions.

Physical factors

Repair materials are generally stronger and more impermeable than the parent concrete so that degradation due to physical factors is not especially critical in a marine situation. The factors listed in section 11.2.4 need to be considered. As the sea temperature changes slowly the marine climate is less extreme than on shore. However, there are parts of the world, such as the Bay of Funday, where each high tide provides a thaw cycle following exposure of the concrete at low tide to intense freezing. Up to 365 freeze–thaw cycles per year can be experienced in these latitudes.

Structural compatibility

Structural compatibility requirements in a marine environment are similar to those for onshore structures.

11.6.3 Research into repair materials and methods for offshore structures

An independent research project in two parts has been conducted by the Building Research Establishment (John *et al.*, 1983; Coote *et al.*, 1986) to test 'the durability of concrete repair systems with special reference to their ability to provide continued protection to the reinforcement steel'. This project is ongoing but the conclusions after six and a half years of exposure are as follows.

1. Rapid-setting systems suffer from long-term dimensional instability.

2. For protection of more than five years ordinary Portland cement (OPC) or OPC–SBR-modified mortars have been shown to be most effective as they provide inhibitive protection which enhances the likelihood for passivity to develop and resist chloride ion migration to the reinforcing steel.

3. Under resin repair systems, steel priming coats need to be inhibitive to avoid crevice corrosion under chloride conditions.

Three commercially based studies, part funded by the Department of Energy and the European Economic Community, have been aimed at developing suitable repair systems for use offshore. These programmes have been carried out by Wimpey Laboratories Limited (Billington, 1979), McAlpine Offshore Limited (Humphreys, 1980; Hill, 1984) and Taylor Woodrow Research Laboratories (CIRIA, 1983). The last of these concentrates mainly on the structural aspects of repairs. The above references mainly refer to details of specimens and test methods but give no results with respect to durability.

The structural implications of repairs to concrete slabs previously damaged by the impact of dropped objects is being investigated under the Concrete Offshore in the Nineties (COIN) programme (Wimpey Laboratories, 1987b).

Some research was carried out on the effects of repairs on the fatigue of reinforcing bars in concrete in the Concrete in the Oceans Research Programme (Leeming, 1989). This work showed that repairing a beam with a cementitious repair material had no adverse effect on the fatigue life of Torbar embedded in concrete. However, it was found that the fatigue endurance of a beam with an epoxy resin repair was generally lower than that of an equivalent standard beam. This is believed to be the result of an alteration of the crack pattern which concentrates the stress in the bar, reducing the extent of the crack blocking and also providing a less effective corrosion protection. However, these results were based on only four tests for each type of repair and exhibited considerable scatter.

Much research carried out on repairs in general has applications to marine situations and has been reported elsewhere in this book.

11.7 Repair strategies and conclusions

The extent to which a structure will need repair will be constrained by the extent to which continued serviceability is required and by the need for structural safety. There are several approaches to repairs which will depend on the requirements of the structure.

1. Do nothing as eventual failure is beyond its intended useful life.

2. Do sufficient to slow deterioration so that the intended useful life is at least met.
3. Accept that repairs have only a finite life and that further repairs on a regular basis will be needed to extend the life of the structure.
4. Accept the present deteriorated state of the structure and stabilize the deterioration at that level for the remaining life of the structure.
5. Restore the structure as closely as possible to its original condition.

The choice of strategy may be different in the two broad categories of marine structures, namely coastal and offshore. The offshore industry takes a more detailed interest in repair methods. Serviceability and safety requirements are relatively well defined and offshore structures are generally massive. Many coastal structures are of beam, slab and column type; their future requirements are less well defined and standards of construction and maintenance may be lower.

The splash zone is the area where structural damage and deterioration as a result of corrosion of the reinforcement is greatest. There are special problems in carrying out repairs to marine structures such as the impossibility of keeping the repair area dry and the contamination from chlorides and marine growth. These problems require special methods and materials. The durability requirements of a repair to a marine structure are different from, and more severe than, those to a building.

There is at present practically no published information on the durability of repairs in marine conditions so that there is no yardstick by which to judge the choice of method or material for repairs. The problems of access and the difficulty of getting the work done in marine conditions make the choice of the right method and material much more important from the economic point of view.

References

ACI (American Concrete Institute) (1980) _Conference on Performance of Concrete in a Marine Environment_, American Concrete Institute Special Publication 65, New Brunswick, August.

Billing, C.J. (1979) Underwater repair of concrete offshore structures, _Offshore Technology Conference, 11th Annual Proceedings_, Houston, TX, April 30–May 3.

BRE (Building Research Establishment) (1982) The durability of steel in concrete; Part 2, Diagnosis and assessment of corrosion-cracked concrete, _Digest 263_, BRE, Garston.

Cavalier, G. and Vassie, PR. (1981) Investigation and repair of reinforcement corrosion in a bridge deck, _Proc. Inst. Civil Eng., Part 1_, **70**, August, pp, 461–80.

CIRIA (Construction Industry Research and Information Association) (1983) Repairing concrete structures, *Offshore Research Focus 40*, CIRIA for the Department of Energy.

CIRIA (1985) *Proceedings of the 1st International Conference on Deterioration and Repair of Reinforced Concrete in the Arabian Gulf*, 26–29 October, Bahrain.

Concrete Society (1984) Repair of concrete damaged by reinforcement corrosion, Report of a Working Party, *Technical Report 26*, Concrete Society.

Coote, A.T., McKenzie, S.G. and Treadaway, K.W.J. (1986) Repairs to reinforced concrete in marine conditions: assessment of a method of studying and evaluating repair systems, *Int. Conf. Marine Concrete 86*, September, Concrete Society, London.

Department of Energy (1984) *Collision of Attendant Vessels with Offshore Installations*: Part I *General description and principal results*; Part II — *Detailed calculations*, Department of Energy Report No OTH 84-208/209, HMSO, London, August.

Faulds, E.C., Philip, M.G. and Neffgen, J.M. (1983) *Operating Experience and Repair Case Histories of Five Offshore Concrete Structures. Design in Offshore Structures*, Thomas Telford, London.

FHA (Federal Highway Administration) (1987), *Protective Systems for New Prestressed and Substructure Concrete*, US Department of Transportation, Report No FWHA/RD-86/193, April.

Glassgold, I.L. (1982) Repair of seawater structures — an overview, *Concrete International*, March, p. 47.

Hill, T.B. (1984) Underwater repair and protection of offshore structures, *Concrete*, **18** (5), 16–17.

Humphreys, B.G. (1980) Underwater repair of concrete, *International Symposium on Behaviour of Offshore Concrete Structures*, October, Brest (France), CNEXO.

ICE (Institution of Civil Engineers) (1977) *Conference on Maintenance of Maritime Structures*, October, London.

ICE (1986) *2nd International Conference on Maintenance of Maritime and Offshore Structures*, 19–20 February, London.

John, D.G., Treadaway, K.W.J. and Dawson, J.L. (1983) The repair of concrete — a laboratory and exposure site investigation, *Corrosion of Reinforcement in Concrete Construction*, Society of Chemical Industries Conference, London, June, Chapter 17, p. 263.

Leeming, M.B. (1989) *Concrete in the Oceans Programme, Co-ordinating Report on the Whole Programme*, Offshore Technology Report OTH 87 248, HMSO, London.

Marine Technology Support Unit (1986) *Review of the Offshore Fire Research Programme*, Offshore Technology Report OTH 86 229, HMSO, London.

UEG (Underwater Engineering Group) (1983) *Repairs to North Sea Offshore Structures — a Review*, Report UR 21, UEG, London.

UEG (1986) *The Influence of Methods and Materials on the Durability of Repairs to Concrete Coastal and Offshore Structures*, UEG Publication UR 36, CIRIA/UEG, London.

Wimpey Laboratories (1987a) *Repair of Major Damage to the Prestressed Concrete Towers of Offshore Structures*, Offshore Technology Report OTH 87/250, HMSO, London.

Wimpey Laboratories (Brown, I.C. and Perry, S.H.) (1987b) *The Assessment of Impact Damage caused by Dropped Objects on Concrete offshore Structures:*

Part A, *Review of Object Types and Velocities and of Available Empirical Formulae*; Part B, *Development of a New Assessment Method*, Offshore Technology Report OTH 87/240, HMSO, London.

Index

Abrasion 255
  resistance 152
Absorption 69
Access 39, 174–6, 178
Acid attack 26–7
Acoustic emission 47
Acrylic
  coatings 148–9, 152, 162
  emulsions 94, 98, 119, 211
Admixtures
  calcium chloride 35, 217
  superplasticisers 13, 104
Aggregates
  heating 32
  marine 35
  porous 31
  reaction 22–3
  petrography 73
  shrinkage 20
  type 67
Air entrainment 8, 31, 231
Alkali-silica reaction 6, 8, 22–3, 146,
    157, 166, 206, 221, 233
Alkyd coatings 149
Anodes 131, 136–7, 141–4, 249

Bills of quantities 179
Bitumen coatings 149
Blast furnace slag 11, 20, 30, 75
Bleeding 16, 74
Bond coats 89, 94–7
  mixing and application 108–9
Bond strength 89, 95–7, 98, 251, 261
Bridges
  highway 188–204
  railway 206–25
Bush hammering 153, 224

Capillary rise 69, 157

Carbonation
  depth 54–5, 76, 146, 157, 176,
    182, 194, 210, 213, 218
  of repair materials 83
  process 7, 24, 33–5, 105, 131, 233
Cathodic protection 93, 130–45, 167,
    196, 198–200, 217, 249
Cement
  chemistry 11–13
  content determination 72, 197
  high alumina 6, 11–12, 13, 21–2,
    166, 246
  hydration 12–13, 18
  low heat 20
  magnesium phosphate 246
  paste 13–15, 67, 74–5
  portland 11, 84
  rapid hardening 246
  shrinkage compensating 99, 103,
    104
  sulphate resisting 6, 30
  type determination 75
Cementitious
  coatings to steel 91–4, 107–8
  grouts 102–3
  levelling coats 100–1
  mortars 83, 85–6, 97–9, 111–16
Check lists 44–6, 175–6
Chemical analysis 70–2, 197
Chloride
  deicing salt 189–92, 202–3, 208,
    215
  depassivation of steel 35–6, 131,
    196–7, 252
  diffusion through repairs 83, 253
  ingress 7, 14, 24, 145, 157–62,
    166–7, 185, 197, 233
Chlorinated rubber 153, 161
Cladding panels 174, 177, 185–6, 219

Coastal structures 258-9
Codes of practice 166-7, 203, 260
Concrete
  chemical analysis 70-2
  constituents 2, 13-15, 203
  curing 99-100, 203
  deterioration 15-36, 37, 177
  historic 1
  petrographic examination 72-6
  physical testing 67-70
  prestressed 37, 131, 140, 188, 190, 200
  quality 51-62, 67-70, 73-4
  reinforced 3, 37, 130, 182, 185, 188, 206
  removal 105-6, 196-8
  replacement 200-1
  spalling 174, 177, 196, 228-30, 235, 252-3
  substrate preparation 97, 100, 118
  thermal conductivity 32
  thermal expansion 18-20
  visual examination 66-7
Conducting paints 142
Construction faults 253
Coring 18, 51-3
Corrosion
  cells 134
  inhibitors 91, 107, 196
Cover meter 46-7
Cracking 6, 8, 16-19, 23, 67, 75-6, 196, 218, 230, 231, 232, 236, 238, 240, 241-2
Crack repairs 101-3, 119-25, 167, 218, 253, 255
Curing of repairs 99-100, 115-16, 119

Demolition 180, 210, 212
Density 68
Desalination 139
Diffusion resistance coefficient 156
Disc cutting 106
Drainage 40
Drawings 39, 176
Dynamic response

Efflorescence 26
Electrochemical corrosion 131-4
Epoxy
  applications 88

bond coats 97, 108-9
concrete coatings 148-9, 152, 153, 161
crack injection 102-3, 119-25, 218, 258
mixing 110-11
mortar 82, 84, 87-9
placing 113-16
steel coatings 92-4
Exotherm 82-3, 87

Fairing coats 100-1
Faraday's law 130, 132
Fillers 85, 86, 99
Fire damage 32-3, 254-5
Flooring compositions 86
Floors 152
Formulators 84-5, 89
Formwork for repairs 114, 116-8
Freeze-thaw 7, 146, 162
Frost attack 25, 30-1, 206, 220-1

Gouges 258
Grit blasting 107, 153, 179, 183, 211, 224

Half-cell potential 62-4, 138, 193, 201-2, 209-10, 217
Heat of hydration 18-9
High-rise buildings 173-87
Highway
  bridges 188-205
  pavements 226-47
Hydrogen gasification 138
Hydrophobic materials 149-50
Hydrostatic pressure 251

Impact damage 249, 252, 254-8
Impressed current 137, 142, 200
Inspection
  contracts 40, 173
  programmes 174-7, 227-8
  staff 175
  visual 39, 66-7
Investigation
  pavements 227-8
  planning 38-40
  structural 169, 174-5, 177, 181
Ionic movement 131-2, 138

Jetties 258-9

Joints
  contraction 20, 226–7
  defects 228–30, 237–9
  expansion 190–1, 203, 226
  inspection 39
  repairs 243–5
  seals 234–5, 242–3
  warping 226

Lime leaching 76
Load testing 47–51
  creep 51
  loading medium 48–9
  monitoring 50–1
  pass criteria 51
  safety supports 49–50

Maintenance 37, 167–9, 180–1, 228
Marine structures 248–65
Microsilica 100
Mixing repair materials 110–3
Mosaic finishes 174, 177
Movement 177

Needle gunning 153
Non-destructive testing
  kits 206
  near to surface 55–8

Ohm's law 132
Oil platforms 141–2, 248, 254–8
Oleoresinous varnish 148–9
Overlays 246–7
Oxygenation 139

Parapets 191
Pavements
  airfield 227
  highway 226–47
Petrographic examination 72–6
Permeability
  in-situ 60–2
  laboratory 69–70, 159
Photography 44
Piers 259
Pigment volume concentration 156
Pipelines 130, 141
Placing
  mortars 113–6
  superfluid microconcretes 116–19
Plastic

settlement 16–18
shrinkage 16–18, 231, 240
Polyester
  coatings 148–9
  mortar 82, 84, 87–9
  placing 113–16
  primers 88
Polyurethane coatings 148–9
Polyvinyl acetate emulsions 94, 98,
  251
Pop-outs 232, 240–1
Property mismatch 89, 167, 198, 261
Porosity 13–14, 68–9
Public safety 177–8, 181, 185, 228
Pulverised fuel ash 11, 30, 75
Pumping 237

Quays 259

Radar 47
Radiography 193
Railway structures 206–25
Reinforcement
  cleaning 106–7
  corrosion 7, 33–6, 130–4, 193,
    207–20, 233
  cover 46–7, 17, 182
  epoxy coated 134, 167, 203
  protection 251–2, 261
  treatments 90–4, 107–8, 178, 182,
    211
Repair
  costs 9, 88, 174, 178, 180, 183,
    192, 194, 204
  durability 250–2, 252, 260–2
  holding 178–9, 210
  materials 84–105
  material testing 83, 260
  monitoring 201–2
  strategies 177–82, 194–201,
    262–3
  techniques 105–27
Resin mortars 86–9
Resistivity 64–5, 132–3, 194, 200
Retaining walls 192
Rust converters 90–1

Sacrificial protection 136
Salt crystallization 31–2
Safety of materials 127–8
Sale of buildings 180

Sampling
  chemical analysis 71–2
  cores 51–3, 65–6, 68
  drilled 54, 66, 197
  lump 65–6
Saponification 149, 251
Sand-cement mortar 82, 85, 98
  polymer modified 98–9, 196
Scaffolds 177–8, 183, 198
Sedimentation 15–18
Shrinkage
  compensation 99, 198
  drying 20–1,226
  of epoxies 87
  of polyesters 87
Silanes 151, 152–3, 157, 161
Silicates 151, 161
Silicofluorides 151
Silicones 149–51
Siloxanes 151, 161, 212
Slant shear test 95
Slip bricks 174
Splash zone 248–9
Sprayed mortar and concrete 83, 217,
    255, 259
  dry process 125–6
  wet process 126
Stearates 151, 161
Stray currents 139–40
Strength
  cementitious mortar 98
  resin mortar 88
  superfluid microconcrete 104
Structural failure 4
Styrene acrylates 156
Styrene butadiene rubber
  coatings 149, 161
  emulsions 94–5, 98, 119, 184, 218,
    251
Submerged structures 139, 249
Sulphate attack 13, 21, 27–30, 166,
    206, 233, 251
Sulphur dioxide 25
Superfluid microconcretes 103–5, 167,
    246
  mixing 113
  placing 116–9

Surface
  appearance 222–4
  scaling 231, 240
  texture 231, 239–40
Surface treatments 146–65, 167
  application 152–3
  coatings 147–9, 224
  control of chloride ingress 157–62
  control of moisture ingress 157–62
  hardeners 151–2
  pore blockers 151–2, 153
  pore liners 149–51
  renderings 152, 153
  sealers 147–9
  thickness 147
Surveys
  carbonation 182
  chloride 185, 209, 217
  covermeter 46–7, 175, 182
  interpretation 76–8
  resistivity 133
  tapping 46, 178, 221
  visual 44–6, 174–5, 193

Tenants 176, 178, 182
Thermal
  coefficient 19, 89
  cracking 19–20, 166
  expansion 19
  stresses 83, 89
Thermography 47

Ultrasonics 58–60
Urea 203

Vacuum impregnation 152
Vapour pressure 157
Video recording 228
Vinyl acetate copolymers 98
Voidage 68

Waterproofing membranes 189–90,
    203, 208
Water jetting 105–6, 153, 197–8
Weathering 15, 24–26
Wire brushing 153

Zinc rich coatings 93–4